Topics in
Current Physics

45

Topics in Current Physics   Founded by Helmut K. V. Lotsch

Structural Phase Transitions II

Edited by K. A. Müller and H. Thomas

With Two Contributions by
K. A. Müller and J. C. Fayet
F. Borsa and A. Rigamonti

With 50 Figures and 47 Panels

Springer-Verlag Berlin Heidelberg GmbH

Professor Dr. h. c. mult. K. Alex Müller

IBM Research Division, Zurich Research Laboratory, CH-8803 Rüschlikon
and University of Zurich, Physics Department, CH-8001 Zurich, Switzerland

Professor Dr. Harry Thomas

Institut für Physik, Universität Basel, Klingelbergstrasse 82
CH-4056 Basel, Switzerland

ISBN 978-3-662-10115-5

Library of Congress Cataloging-in-Publication Data. (Revised for vol. 2) Structural phase transitions. (Topics in current physics ; v. 23,) Includes bibliographies and indexes. 1. Phase transformations (Statistical physics) 2. Solid state physics. 3. Order-disorder models. I. Müller, K. A. (Karl A.), 1927– . II. Thomas, H. (Harry) III. Dorner, B. (Bruno) IV. Series. QC176.8.P45S77 1981 530.4′1 80-23544
ISBN 978-3-662-10115-5   ISBN 978-3-662-10113-1 (eBook)
DOI 10.1007/978-3-662-10113-1

© Springer-Verlag Berlin Heidelberg 1991
Originally published by Springer-Verlag Berlin Heidelberg New York in 1991
Softcover reprint of the hardcover 1st edition 1991

2154/3140-543210 – Printed on acid-free paper

Preface

In the series *Topics in Current Physics*, this is Volume II on structural phase transitions. It is a continuation of Volume I with series number 23, and, like Volume I, is devoted to experiments. However, it is clearly distinct from it. Volume I contains three chapters: on optical studies, inelastic neutron scattering and ultrasonic investigations, all of which probe collective excitations of the lattice. The present volume has two chapters, the first on electron paramagnetic resonance (EPR) and the second on nuclear magnetic and nuclear quadrupolar resonance (NMR–NQR). These techniques probe local properties and are relevant to the understanding of the phenomena in the field.

The methods of investigation in structural phase transitions (SPT) range widely, so that, on initiating this group of books, it was decided that knowledgeable scientists working with a particular technique should elucidate their method and its characteristic results as well as its advantages and limitations. As could be expected, each chapter of Volume I varied in style, but they were always kept such as to be comprehensible to the nonspecialist. This is also the case for the contributions in the present volume, but even more so as the presentation in the two chapters is really different. The first, by K.A. Müller and J.C. Fayet on EPR, is a review in the traditional style, whereas that by F. Borsa and A. Rigamonti uses a new didactic approach: In so-called panels containing figures and the relevant analytical equations, the essence of a phenomenon is condensed. The rationale for this approach is given hereafter.

More importantly, the emphasis of the two chapters is rather different. That on EPR is more heavily weighted towards static results, that on NMR more towards dynamics. This is also the main reason for their order: EPR has made substantial contributions to studies of static properties of SPT with high sensitivity. Thus, the first two-thirds of the EPR chapter reviews these results, ranging from mean-field behavior to critical and multicritical phenomena. The last part leads on to dynamical aspects, especially in order-disorder systems, and shows the remarkable success of EPR in incommensurate structures. Because of the importance of the method and requests from colleagues working in other fields for an introduction to EPR, considerable information on the methodology is given at the outset.

NMR and NQR are powerful probes for studying solids and especially SPT in an intrinsic way, although less sensitive than EPR. However, what one really probes by the various NMR line splittings and relaxation times is even more remote than EPR to the non-resonance scientist, and this is a major reason why the authors adopt the presentation in the form of panels. In addition, NMR has yielded very interesting recent results on nonlinear dynamics, incommensurate transitions, disordered systems and the central peak, so that a compact type of presentation was also needed to keep the chapter to a manageable size.

Volume II appears a considerable time after Volume I. However, this is not a disadvantage as such, because new aspects in the SPT fields have been included, as mentioned above for the NMR chapter, as well as the recent analysis of extrinsic vs intrinsic properties of EPR parameters made possible by the superposition model. The latter were only elucidated in a conclusive way in 1986. Also, the important EPR results on incommensurate systems are very recent. The intended completion of this volume by 1987 would have been possible; however, in 1986 high-T_c superconductors were discovered. As one of the editors and authors was directly involved in this discovery, he was no longer master of his time regarding other commitments.

It was originally planned to include a third chapter on calorimetric and dielectric properties in the present book. A first draft was reviewed, but the final version was not available when printing was initiated. As the two chapters on resonance are already quite sizeable and together form a true unit addressed to local probing, it was felt more appropriate to include the review on calorimetry in the planned Volume III on theory, which will also contain a chapter on thermodynamics. Of course, calorimetry and dielectric behavior are closely related to the results in thermodynamics.

The introduction which appeared in Volume I was written with the intention of serving for all experimental chapters, including the present two. Its outline is still up to date and should be of help in opening the door to SPT for those interested. Furthermore, each of the present chapters begins with an introduction, and thus it was felt that a separate introduction to this book would be superfluous. The editors hope that the present effort will serve to familiarize researchers and students with magnetic resonance techniques as applied to SPT. It may be pointed out that since the proceedings of the Enrico Fermi Varenna School on "Local Properties of Structural Phase Transitions", which was published in 1976, no comparable presentation on EPR and NMR studies of SPT has appeared in the literature.

Rüschlikon and Basel,
January 1990

K.A. Müller, H. Thomas

Contents

List of Contributors

Ferdinando Borsa
Università di Pavia, Dipartimento di Fisica, "A. Volta",
Unità INFM-GNSM and Sezione INFN, Via A. Bassi 6, I-27100 Pavia, Italy

Jean Claude Fayet
Laboratoire de Spectroscopie du Solide, E.R.A., Faculté des Sciences,
F-72017 Le Mans Cedex, France

K. Alex Müller
IBM Research Division, Zurich Research Laboratory, CH-8803 Rüschlikon and
University of Zurich, Physics Department, CH-8001 Zurich, Switzerland

Attilio Rigamonti
Università di Pavia, Dipartimento di Fisica, "A. Volta",
Unità INFM-GNSM and Sezione INFN, Via A. Bassi 6, I-27100 Pavia, Italy

1. Structural Phase Transitions Studied by Electron Paramagnetic Resonance

K.A. Müller and J.C. Fayet

With 50 Figures

1.1 Introduction

Electron Paramagnetic Resonance (EPR) has been one of the pioneering techniques used for detecting Structural Phase Transitions (SPT) and determining the space groups involved and especially for investigating, with high precision, the order parameter and its dependence on temperature and stress. Earlier work on SPT was reviewed by one of us [1.1] and summarized for the model substances $SrTiO_3$ and $LaAlO_3$ over a decade ago by *Müller* and *Von Waldkirch* [1.2]. In the present review, we attempt to cover a broader field, including the most recent progress in EPR research in incommensurate SPT's and random systems.

Since the invention of the laser and the availability of intense neutron sources, a great deal of solid-state work has been carried out by scattering methods with the result that EPR, previously taught in most solid-state departments, is now accorded less emphasis. [However, it is still widely used along with Nuclear Magnetic Resonance (NMR)]. We have thus decided to review this method in Sect. 1.2 in some depth, for the benefit of those readers who are not familiar with it, or for those who wish to refresh their memories. For a thorough study we recommend, among many good books, the one by *Pake* and *Estle* [1.3], entitled *Principles of Paramagnetic Resonance*, from which university courses have often been taught and where the material presented is of adequate sequence and length. *Abragam* and *Bleaney*'s monumental treatise [1.4] is more detailed, and was useful as a general reference up to the midsixties when most of the fundamental aspects of EPR had been settled. To complete the picture, the new edition of the book by *Altshuler* and *Kozyrev* [1.5] gives more access to the Russian literature. For electron spin resonance of radicals, we would also like to mention the book by *Atherton* [1.6].

In Sects. 1.2 and 1.3 we further emphasize aspects of EPR that are specific to the study of SPT, namely the paramagnetic ions to be chosen, the use of the superposition model, and the additional equipment necessary to employ this inexpensive method successfully. Particularly important is the recent progress due to the use of the superposition model. The subsequent sections form the heart of our article and present characteristic examples. Because of the vast field covered by our review, we present highlights of the method rather than a complete study. Oxide ferroelectrics are presented in Sect. 1.4, antiferrodistortive transitions in Sect. 1.5, multicritical points in Sect. 1.6, EPR in order-disorder transitions in Sect. 1.7, and work on incommensurate systems in Sect. 1.8.

1.2 Methodological Aspects

In this section, we review the concepts and techniques of electron paramagnetic resonance for investigating structural phase transitions in solids. EPR in condensed matter was discovered in 1945 by *Zavoisky* [1.7] as the first of a group of phenomena which comprise nuclear magnetic resonance (NMR), ferromagnetic and antiferromagnetic resonance, cyclotron resonances, and other techniques such as various kinds of double resonance. It is observed when a high-frequency magnetic field induces transitions between the Zeeman levels of a dilute magnetic defect in a nonmagnetic solid or liquid. The magnetic levels E_M depend on the orientation of an applied external field H with respect to the local crystal or ligand field of the ion. This interaction is a direct one for the orbital states, whereas for spin states it occurs through the spin-orbit interaction

$$\mathcal{H}_{SO} = \sum_i \zeta s_i l_i \quad ; \quad l_i = r \times p_i \tag{1.1}$$

and the spin-spin interaction

$$\mathcal{H}_{SS} = \sum_{ij} (2.0023 \times \mu_\beta)^2 \left[\left(\frac{(s_i s_j)}{r^3} - 3 \frac{(s_i r_i)(s_j r_j)}{r^5} \right) \right] \quad , \tag{1.2}$$

which couples spin s_i to space coordinates r_l; the single-electron spin-orbit constant and the Bohr magneton being ζ and μ_β, respectively [1.2–4, 8]. The orbits l_i in $\mathcal{H}_{SO}$ are influenced by the local crystal field, and the distribution of electronic density resulting from the latter changes $\mathcal{H}_{SS}$. Therefore, the EPR transition measured at a given microwave frequency ν,

$$h\nu = E_M - E_{M-1},$$

reflects the point symmetry of the crystal field to which an impurity ion is exposed. We shall emphasize the choice of ions for such studies and the spin-Hamiltonian formalism to describe the spectra, especially the importance of the various terms which reflect the *local symmetry*. They can be used to determine or confirm the space group of a crystal, to observe any structural transformations that might occur, and to determine the order parameter and its variation with temperature and external fields. In displacive systems, the order parameter can be defined as a local parameter $\eta(x)$ [1.2]. EPR measures it most directly. In order-disorder systems, the order parameter is proportional to the differences of occupation numbers. Measurement of this quantity is one of the most recent successes of the EPR method.

In order to draw conclusions about the crystal symmetry, it is necessary to know where the impurity ion is located. We shall give criteria for determination of this site in Sect. 1.2.3. The salient features of the experimental procedure are then outlined together with some remarks on the sensitivity and limitations of the method.

1.2.1 Choice of Paramagnetic Ions or Radicals

The essence of the method consists in substituting or creating a paramagnetic ion or radical in a nonmagnetic crystal. To probe the local crystal-field symmetry at the ion site, the splittings of the spin levels E_M are investigated. In order that they reflect the point symmetry of the virgin crystal, it is important that the magnetic substituent matches the nonmagnetic species as closely as possible. Thus it should have the same size, effective valence, and a half-filled (nonbonding) subshell. While the first two requirements are obvious, the latter is important for a number of reasons: even in crystals with a certain covalency the nonmagnetic host ions or atoms have almost spherical electronic shape. This property is best matched by a paramagnetic ion with a half-filled d or f subshell. Such ions have necessarily nondegenerate orbital ground states, since the total orbital angular momentum is $L = \sum_i l_i = 0$ from Unsöld's theorem. This is true regardless of the sign of the crystal field. Ions with $L \neq 0$ can lower the observed local symmetry due to linear coupling with the local coordinate as a result of the Jahn-Teller effect [1.3, 4], if the ground-state orbital moment has not already been quenched by the crystal field. But even if it has already been quenched, unfilled d or f subshells, other than half-filled ones, can be dangerous owing to their nonspherical spacial charge density! An example of this is the Cr^{5+} ($3d^1$) ion replacing the As^{5+} in KH_2AsO_4 [1.9]. The charge and size match perfectly, but the paramagnetic orbital has $d(x^2 - y^2)$ symmetry. Each of the x^2 and y^2 lobes couples with two protons in lateral Slater configurations forming a $Cr^{5+}O_4H_2$ unit which can re-orient. Thus, the local point symmetry is lower than that of the host.

We therefore recommend the use of ions with half-filled shells in order to determine lattice cation point symmetries. For the $3d$ and $4f$ shells, these are

$$3d^5 : Cr^+ , Mn^{2+}, Fe^{3+} \quad S = \tfrac{5}{2} \quad \text{or}$$

$$4f^7 : Eu^{2+} , Gd^{3+}, Tb^{4+} \quad S = \tfrac{7}{2} \; ,$$

the choice depending on the size and valence of the substituted cation. Further advantages in using such ions are:

- The half-filled subshells of these ions always have a high energy of the first excited state [1.4]. This, together with the nearly vanishing orbital angular momentum contribution in the ground state, implies long spin-lattice relaxation times at high temperatures owing to the weak spin coupling. The relaxation times in the liquid-helium temperature range are also reasonable. Therefore studies are possible from low temperatures up to $1000\,K$ and higher.

- They are best suited because they have spins $S = \sum s_i$ larger than three halves ($S > \tfrac{3}{2}$). Therefore, they show splittings of hexadecapolar or higher form [1.3, 4], i.e., terms of $S_x^4 + S_y^4 + S_z^4$ or higher in the Hamiltonian (Sect. 1.2.2). For symmetry purposes, this makes the EPR method superior to nuclear magnetic resonance and Mössbauer studies. In the latter, only quadrupolar splittings have so far been observed. Thus, only the orientations of the *axial* crystal fields can be obtained, whereas EPR also allows one to probe cubic field orientations, which may be *all important* for the determination of local symmetry axes in certain cases.

- Except for the very ionic crystals, the second-order ligand-field term results from the nearest-neighbor ions [1.10]. They can be taken into account by the superposition model reviewed in Sect. 1.2.4. This sensitivity of EPR to nearest-neighbor charge distribution is in contrast to the NMR method, which also probes local properties by the nuclear quadrupole splittings (NQS) [1.11]. The latter depend considerably on atomic monopole and dipole contributions which converge slowly and make NQS dependent on distant atoms.
- In general, for half-integral effective spin ground states (which include those having half-filled shells), all EPR lines shift in first-order perturbation on lowering the symmetry except the purely magnetic $M_s = \frac{1}{2} \rightarrow -\frac{1}{2}$ line. This offers an advantage over normal diffraction methods (X-rays, neutron diffraction, etc.) because in these methods, in the most favorable case, only special reflection spots split or become permissible.

The last of these advantages pertains to all paramagnetic ions and radicals even if the local symmetry differs from that of the host, and has thus frequently been employed to detect a structural phase transformation. Moreover, certain lattice defects are sometimes more sensitive to changes in order parameters than those which do not disturb the local point symmetry. The Fe^{3+}-V_O pair defect in $SrTiO_3$ [1.12] and the Gd^{3+}-O^{2-} center in $RbCaF_3$ [1.13] are typical examples, when compared to the $Fe^{3+}O_6$ or $Gd^{3+}F_6$ complexes, respectively. The procedure here is first to establish the linear or quadratic dependence of the EPR line splittings and/or shifts of the low point-symmetry defect on the order parameter, and compare them to EPR results on non-symmetry-disturbing defects or to the order-parameter dependences obtained by other methods, e.g., NMR, scattering, birefringence, etc. Then, the dependence established is used for the intended investigation. In electron spin resonance (ESR) work, this is the predominant way of using radicals which, in one way or another, do not usually match the microscopic configurations of the host. In ambiguous cases, it is advantageous to compare experiments carried out with two or three impurities to ensure that the results are consistent.

The use of EPR data for *different impurities* situated at the *same site* has most recently further extended the advantages of the method. As will be seen in Sect. 1.2.4 on the superposition model, two ions of the same valency and size can probe different properties of their environments. We shall summarize there the sensitivity of the Fe^{3+} and Cr^{3+} ions, both of the same size, with respect to their nearest neighbors. The second-order ligand-field parameters of the former are very sensitive to the *distance* from their nearest neighbors, whereas those of Cr^{3+} are not all, but are highly sensitive to the *angle* formed between Cr^{3+} and the ligands [1.14]. Before proceeding with this subject, an introduction to the formalism with which the EPR data are analyzed is necessary.

1.2.2 The Spin-Hamiltonian Formalism

The spin Hamiltonian is the traditional way of representing the information furnished by the electron paramagnetic resonance [1.3–5]. This formalism describes the splitting of the ground-state levels of a paramagnetic ion as an expansion in spin operators S_z, $S_x = \frac{1}{2}(S_+ + S_-)$, $S_y = (-\frac{i}{2})(S_+ - S_-)$ in a base of $|S, M\rangle$ functions. The latter are eigenfunctions of the S^2 and S_z operators. In the case of half-filled shells, the orbital angular momentum $L = 0$, and $S = \sum_i s_i$ is a *true* spin Hamiltonian. However, normally S is just used to describe the multiplicity of the ground state, and therefore the spin Hamiltonian is an *effective* one. Up to fourth order in the S-operators, the Hamiltonian is then given by

$$\mathcal{H} = \sum_{ij} \beta H_i g_{ij} S_j + \sum_{m=-2}^{m=+2} B_2^m O_2^m + \sum_{m=-4}^{m=+4} B_4^m O_4^m \quad . \tag{1.3}$$

For $S = \frac{7}{2}$ (Gd^{3+}), small terms of sixth order, $\sum_{m=-6}^{m=+6} B_6^n O_6^m$, have to be added. The B_l^m are appropriate parameters, the O_l^m are normalized spin operators (see below). In the most general case, together with the g_{ij}, there are 20 constants to be determined for $S = \frac{5}{2}$.

If the interactions (1.1) and (1.2) of the spin with the lattice are switched off, the interaction of the magnetic moment of the spin $\mu = -g\beta S$ with the magnetic field H is

$$\mathcal{H}_z = -\mu H = g\mu_\beta S H \quad . \tag{1.4}$$

Magnetic "resonance" occurs when the energy difference between two consecutive magnetic levels $g\mu_\beta H$ equals the applied microwave quanta $h\nu$. With the inter-actions (1.1) and (1.2) present, the free spin g-value = 2.0023 is changed and can be put into the form of a symmetric tensor $S_i g_{ij} H_j$. Derivations using the ionic approximation have been given [1.4, 11] for half-filled shells; g_{ij} is usually still approximately isotropic and close to 2. Furthermore, at $H = 0$ the levels are split owing to the $s_i r_l$ interactions (1.1) and (1.2). These splittings are represented in the Hamiltonian by terms of the form $S_x^p S_y^q S_z^r$ where time-inversion symmetry requires $n = p + q + r$ to be even. The expansion has to be broken off at $n = 2S$ in order to stay within the $|S, M\rangle$ basis manifold. The terms in powers of $H \times S_x^p S_y^q S_z^r$ with n odd (higher-order Zeeman terms), so far, could not be detected for half-filled shell ions, and have thus been omitted in (1.3). Apart from the Zeeman term (1.4), the Hamiltonian contains terms in $S_x^p S_y^q S_z^r$ grouped together in normalized spin opera-tors $O_l^m (S_x, S_y, S_z)$ which transform as the corresponding homogeneous Cartesian polynomials. The latter are defined as linear combinations of Wigner's tensor oper-ators T_l^m which transform as spherical polynomials Y_l^m (i.e., result from a unitary transformation),

$$O_l^{\pm m} + \left[T_l^{-m} \pm (-1)|M|T_l^m\right] \quad ,$$
$$O_l^0 = T_l^0 \quad . \tag{1.5}$$

For example, $O_2^2 = 3S_z^2 - S(S+1)$, $O_2^2 = (S_x^2 - S_y^2)$.

We now discuss the terms appearing in the Hamiltonian of (1.3). They have to constitute a basis for the irreducible representation Γ_1 of the Hamiltonian $\mathcal{H}$ within the point group of the ion site. For high symmetry, this considerably reduces the number of constants. Consider for instance an Fe^{3+} ($S = \frac{5}{2}$) ion in cubic O_h

symmetry, as found in SrTiO$_3$ above its phase transition [1.15]. The spin Hamiltonian (1.3) will then reduce to

$$\mathcal{H} = g\beta \boldsymbol{H}\boldsymbol{S} + B_4^0\left(O_4^0 + \sqrt{\tfrac{5}{7}}O_4^4\right)$$

$$\equiv g\beta \boldsymbol{H}\boldsymbol{S} + \tfrac{a}{6}\left[S_x^4 + S_y^4 + S_z^4 - \tfrac{1}{5}S(S+1)(3S^2 + 3S - 1)\right] \quad (1.6)$$

if the principal axes are chosen parallel to the x, y, z quaternary $\langle 100 \rangle$ axes. There are now only two constants to be determined: g and $a = 15B_4^0$. The splittings of the M_s levels for H parallel to a $\langle 100 \rangle$ axis (as originally computed by *Kittel* and *Luttinger* [1.16]) are shown in Fig. 1.1a. Because in the resonance experiment the microwave quantum is usually fixed and the magnetic field H is scanned, absorption is observed when $h\nu = E_M - E_{M+1}$. The five allowed $\Delta M = 1$ fine-structure transitions are shown in Fig. 1.1b, together with their intensities which are proportional to $\langle S, M | S_+ | S, M - 1 \rangle^2$. Rotating H away from [100] changes the energy levels arising from the diagonalization of (1.6). Consequently, the ΔM resonance magnetic fields H_n also vary. This is shown in Fig. 1.1c for a variation of H in a (100) plane of the crystal [1.15].

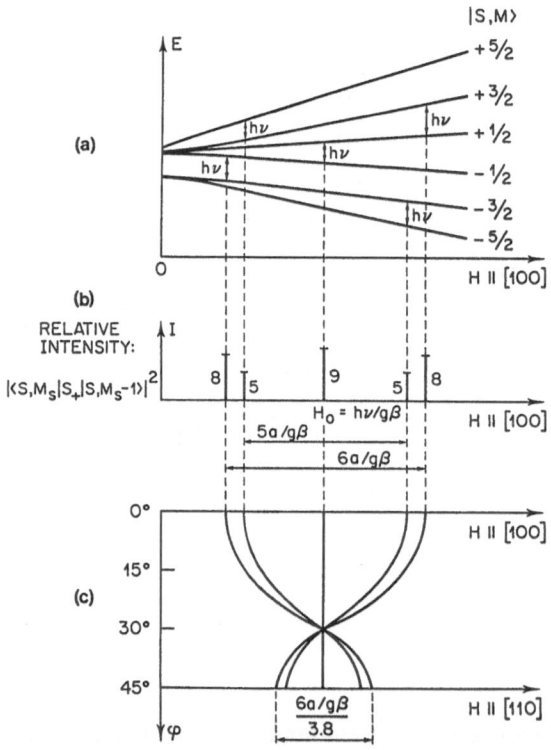

Fig. 1.1. (a) The splitting of an $S = \tfrac{5}{2}$ state in a cubic crystal field as a function of a magnetic field H parallel to a [100] direction. (b) Relative intensities of the allowed $\Delta M = 1$ transitions shown in (a). (c) Anisotropy of the $\Delta M = 1$ lines on rotation of H in a (001) crystal plane [1.15]

In low-symmetry crystals, the B_2^m are often between 10^{-1} to $10^{-2}\,\mathrm{cm}^{-1}$, whereas the B_4^m are about an order of magnitude smaller. The determination starts by assuming g_{ij} to be scalar, and finding the principal axes of the second-order terms. For the latter, only two such terms, $B_2^0 O_2^0$ and $B_2^2 O_2^2$, do not vanish. With these "B_2^m"-type axes fixed with respect to those of the crystal, the fourth-order terms and g_{ij} are obtained by an iteration procedure using the full diagonalization of the spin Hamiltonian matrix as carried out by *Michoulier* [1.17]. He uses resonance magnetic fields with H in sufficiently many directions that the constants g_{ij}, B_l^m are largely overdetermined. The second-order parameters depend on second derivatives of the crystal-field potential, i.e., on local covalency contributions of nearest neighbors as well as on more distant atomic configurations [1.4, 18]. Generally, owing to the different possible origins, these parameters are not entirely reliable for symmetry determinations. This pertains to highly ionic crystals like certain fluorides, where crystal fields contribute a sizable fraction to the B_2^m values [1.19]. However, when a slight covalency exists, the nearest neighbors determine these second-order parameters. This is certainly the case in oxides.

In the ionic models for $3d^5$ ions (partially applicable to Mn^{2+}), the fourth-order terms depend on the fourth-order derivative of the crystal-field potential [1.20], i.e., essentially on *local* electric charge distribution. This will be enhanced if covalency becomes important, as for Fe^{3+}; it is then the location and type of the nearest-neighbor ligands and charges that determine a. The x, y, z directions in (1.3) thus reflect short-range symmetry and, for sixfold coordination, point along the slightly deformed octahedral corners even if the crystal has an appreciably lower overall symmetry. Experimentally, this was first verified for Fe^{3+} in Al_2O_3 [1.21] and $LaAlO_3$ [1.22] both having trigonal $R\bar{3}c$ structure. The rotation of oxygen octahedra around the $\bar{3}$ axes obtained by EPR agreed with those found from diffraction work. Al_2O_3 and $LaAlO_3$ still have quite high symmetry, but later *Michoulier* and *Gaite* [1.17] introduced a generalized procedure for low-symmetry crystals by means of which local pseudo-cubic quaternary Q and trigonal T axes can be determined from a knowledge of the B_4^m parameters. In the feldspath albite $NaAlSi_3O_8$, a triclinic mineral, they showed the following: the local Fe^{3+} EPR axes of quaternary and trigonal symmetry, determined by using their method, agreed with those from diffraction studies for an experimental accuracy of one degree achieved with both techniques. In this mineral, the Fe^{3+} replaces the Al^{3+} ions in distorted tetrahedral oxygen sites. Their result shows that the EPR method can also be employed for very complicated and low-symmetry crystals to determine local pseudo-symmetries possibly not obtainable by diffraction, or alternatively to discriminate between several possibilities left open.

1.2.3 Site Determination

Knowledge of the site at which the impurity ion is incorporated is essential for reaching correct conclusions about the kind of space group to which the crystal belongs or into which it transforms. In structures with large unit cells, one may have to choose between several substitutional or interstitial positions. Often, considerations of crystal chemistry, valence and size are useful as are known symmetry properties from diffraction work. The question arises of whether it is possible to establish, merely by using EPR, where the paramagnetic probe ion is located.

The most frequently-used property to determine the coordination, i.e, whether the ion is at a tetrahedral, octahedral or other site, is the magnitude of the cubic splitting parameter a. As a rule of thumb, for $3d^5$ ions in oxides located at tetrahedral sites, a is about half that of octahedral sites. The value of a for Fe^{3+} is about $2.0 \pm 0.1 \times 10^{-2}\,cm^{-1}$ in the octahedral sites of TiO_2, $SrTiO_3$, MgO, Al_2O_3, etc. [1.23]. Theoretical calculations for $3d^5$ ions using an ionic approach have been made, but we cannot dwell upon them here [1.20].

Another very useful piece of information is the hyperfine interaction. Most paramagnetic ions have at least one isotope with nonzero nuclear moment. The central hyperfine interaction parameter $A = \frac{1}{3}\mathrm{Tr}\{A_{ij}\}$ in the Hamiltonian $\mathcal{H}_{Hf} = \sum_{ij=1}^{3} S_j A_{ij} I_j$ is almost a constant for a particular coordination and ligand, independent of the lattice spacing. For example, for Mn^{2+} octahedrally coordinated in MgO, CaO or SrO, $A \simeq 81.7 \times 10^{-4}\,cm^{-1}$ [1.24]. If the ligand has a nonzero nuclear moment like F, Cl, Br, S, etc., the super-hyperfine interaction (SHF) [1.3, 4] between the electronic spin of the metal ion and the nuclear spin of the ligands yields the coordination and the distortion from high symmetry of the complex.

When the local point symmetry lacks inversion, linear shifts of the lines as a function of an external electric field E can be observed [1.25]. It is then also possible to distinguish between inequivalent sites. The terms in the Hamiltonian are of the form $E_i T_{ijk} H_j S_k$ and $E_i R_{ijk} S_j S_k$ for the shifts in the Zeeman and quadratic crystal-field terms, respectively. E_i are the components of the electric field, and the T_{ijk}, R_{ijk} are constants which are tabulated for various point symmetries together with measured values reviewed by *Mims* [1.26]. He also considers the properties of higher-order terms. In the presence of inversion symmetry, the effects are quadratic and minute, of the order of milliGauss per kV/cm only [1.27]. They are enhanced in high-dielectric-constant materials. For Fe^{3+} in the cubic $KTaO_3$ at 4.2 K, they are about 2 G per kV/cm [1.28].

Application of uniaxial stress σ_{ij} is a further possibility to distinguish between inequivalent sites [1.3]. The additional Zeeman and quadratic-spin terms in the Hamiltonian are of the form

$$R_{ijkl}\sigma_{ij}S_k S_l \quad \text{and} \quad P_{ijkl}\sigma_{ij}H_k S_l \quad .$$

For cubic crystals, only two terms differ from zero: linear combinations of operators S_i^2 multiplied by $R_1 = (\frac{3}{4})C_{11}$ and linear combinations of $S_i S_j\,(i \neq j)$ multiplied by $R_2 = C_{44}$ (C_{nk} are the elastic constants). The properties of the R_{ijkl} and P_{ijkl} constants and measurements thereof have also been reviewed [1.29].

1.2.4 Ligand-Field Parameters and the Superposition Model

Symmetry and structural information is contained in the orientation of the $i = x, y, z$ axes of the $O_n^m(S_i)$ operators of (1.5) as well as in the B_n^m parameters. As an example, we discussed the principal axes of the term $B_4^0(O_4^0 + \sqrt{\frac{5}{7}}O_4^4)$ in cubic symmetry. Near a phase transition occurring at T_c, both the orientation $Q_i(T)$ of the axes x, y, z and the magnitude of the $B_n^m(T)$ parameter can be written as a Taylor expansion of the order parameter η or of one of its components η_k:

$$\left.\begin{array}{r} Q_1(T) - Q_i(T_c) \\ B_n^m(T) - B_n^m(T_c) \end{array}\right\} = a\eta_k(T) + b\eta_k^2(T) + \text{higher-order terms} \quad . \tag{1.7}$$

Usually, the external magnetic field can be chosen in such a way that by symmetry the coefficient a or b is zero, i.e., via (1.3) one detects a linear or quadratic response of the EPR lines to the order parameter or one of its components. The quantitative sensitivity of the response depends strongly on the structural sites in the crystal, and is very important for the experiment.

Ab initio calculations of the spin-Hamiltonian axes and paramters B_n^m based on the crystal field and covalency contributions have been undertaken beginning in the early days when the spin Hamiltonian was first introduced [1.3–5] until most recently [1.19]. These calculations suffer from the existence of large terms of opposite sign which contribute to a particular B_n^m, not yet allowing its quantitative determination [1.19]. On the other hand, the so-called superposition model has given quite valuable results for the second-order [1.10, 14], and for the fourth-order B_n^m parameters for $^8S_{7/2}$ ground-state ions [1.30]. In the presence of a few percent of covalency, the superposition model allows the determination of the spin-Hamiltonian parameters for a known structure of a defect in a crystal, or conversely the deduction of a local, possibly nonintrinsic structure from EPR data. The model was originally introduced by *Newman* and *Urban* [1.10] for the rare-earth $^8S_{7/2}$-state ions Gd^{3+} and Eu^{2+}. The main assumption of the model is that the spin-Hamiltonian parameters result from individual contributions of each nearest neighbor of the paramagnetic ion. Essentially, this means that the interaction resulting from overlap and covalency mechanisms dominates. Because of this assumption, the model is not applicable in strongly ionic compounds where crystal-field calculations, such as the polarizable point charge model, are more appropriate [1.31, 32].

We restrict ourselves here to the B_2^m terms of (1.3) for which the model has been mainly employed, and refer the reader to the literature regarding the magnitude of the B_4^m terms [1.10]. We recall that the Hamiltonian (1.3) can always be transformed to such an axis that only the O_2^0 and O_2^2 terms do not vanish [1.3–5, 10]. The two parameters B_2^0 and B_2^2 multiplying O_2^0 and O_2^2, respectively, are given in the model with the newer notation $b_2^m = 3B_2^m$ by

$$\begin{aligned} b_2^0 &= \bar{b}_2(R_0)\tfrac{3}{2} \sum_1^n \left(\frac{R_0}{R_i}\right)^{t_2} \left[\cos^2 \Theta_i - \tfrac{1}{3}\right] \quad ; \\ b_2^2 &= \bar{b}_2(R_0)\tfrac{3}{2} \sum_1^n \left(\frac{R_0}{R_i}\right)^{t_2} \left[\sin^2 \Theta_i \cos 2\psi\right] \quad . \end{aligned} \tag{1.8}$$

Here, R_0 is a reference distance chosen near the distances R_i between the paramagnetic ion and the ith ligand. Θ_i is the angle between the line joining the paramagnetic ion to the ith ligand and the main axis of the spin Hamiltonian z, and ψ_i is the angle between this axis and the projection of the ith ligand coordinate onto the x, y plane.

In (1.8), it is assumed that $\bar{b}_2(R)$ varies exponentially with R with exponent t_2. The two parameters $\bar{b}_2(R_0)$ and t_2 appearing in (1.8) were determined in two quite different ways: historically first was the method using measured b_2^m's and assuming that the central paramagnetic ion and the ligands occupy intrinsic lattice positions as determined for the bulk crystal, for example, by the X-ray refinement method [1.10, 33]. The use of information from ligand super-hyperfine data [1.34, 35] is another possibility. The second method was introduced later. One employs a cubic crystal such as MgO or $SrTiO_3$, and obtains the parameters by applying stress. The first constants obtained in this manner were the $\bar{b}_4(R_0)$ and t_4 parameters from hydrostatic-pressure data [1.36]. Then, $\bar{b}_2(R_0)$ and t_2 were computed from uniaxial stress-coupling parameters of Fe^{3+} and Mn^{2+} [1.36, 37]. The relations between the strain coefficients G_{11} and G_{44} of the crystal and the two parameters $\bar{b}_2$ and t_z are given by [1.30, 36]

$$G_{11} = -\tfrac{4}{3} t_2 \bar{b}_2 \quad ; \qquad G_{44} = \tfrac{1}{2} G_{5g} = 2\bar{b}_2 \quad , \tag{1.9}$$

thus, $t_2 = -\tfrac{3}{2} G_{11}/G_{44}$. In (1.9), $G_{11} = (c_{11} - c_{12}) C_{11}$ and $G_{44} = c_{44} C_{44}$, the small c_{ij} being the elastic stress parameters of the cubic crystal [1.3].

A comparison of superposition-model parameters for Gd^{3+} in eight-fold, and for Fe^{3+} and Cr^{3+} in octahedral coordination show a ligand coordination number dependence [1.33], a limitation of the model that results from an aspherical charge distribution due to the presence of covalency. For Gd^{3+}, experimental t_2 values close to zero have been accounted for [1.10] with a two-term power law of the form

$$\bar{b}_2(R) = (-A + B)\left(\frac{R_0}{R}\right)^{t_2} = -A\left(\frac{R_0}{R}\right)^n + B\left(\frac{R_0}{R}\right)^m \quad , \tag{1.10}$$

with $(B - A) = \bar{b}_2(R_0)$ and $m > n > 0$. Theoretically, many positive and negative contributions to $\bar{b}_2(R)$ are present, but have been collected in the two terms in (1.10). A difference of $m - n = 3$ has been taken in the Lennard-Jones-type function for Gd^{3+} with $n = 7$ and $m = 10$. For Fe^{3+} with $t_2 \geq 8 \pm 1$, $n = 10$ and $m = 13$ were used [1.14]. This choice of n and m yields $t_2 = 8 \pm 1$ over a reasonable interval around $R_0 = 2.1$ Å, as shown in Fig. 1.2 ($t_2 = 8 \pm 1$ is normally used for all superposition-model analyses of Fe^{3+}). Taking $n = 9$ or $n = 11$ did not substantially alter the results (less than 10 % of $\bar{b}_2$ between $R = 1.9$ and 2.2 Å); a fact well known for Lennard-Jones-type potentials with high exponents.

The binding of transition-metal ions to their neighboring atoms is intrinsically more covalent than that of rare-earth ions for a given kind of ligand. Consequently, the model applied to the S-state transition-metal ions Fe^{3+} and Mn^{2+} has been quite successful [1.36]. The structure of transition-metal oxygen-vacancy pair centers $(Me-V_O)$ in oxides was determined with a high degree of confidence [1.37]. Also the $FeOF_5$ center in $KMgF_3$, Fe^{3+} in $Sr_2Mb_2O_7$ [1.38] and the Mn^{2+} F-center pair in CaO [1.35], as well as the Fe^{3+} in $Na_5Al_3F_{14}$ [1.34] have been accounted for.

In principle, the model is applicable only to ions with half-filled d or f shells having A_1 ground states. The observed coordination-number dependence [1.33] restricts its generality, however. Thus, with the coordination-number restriction, it may

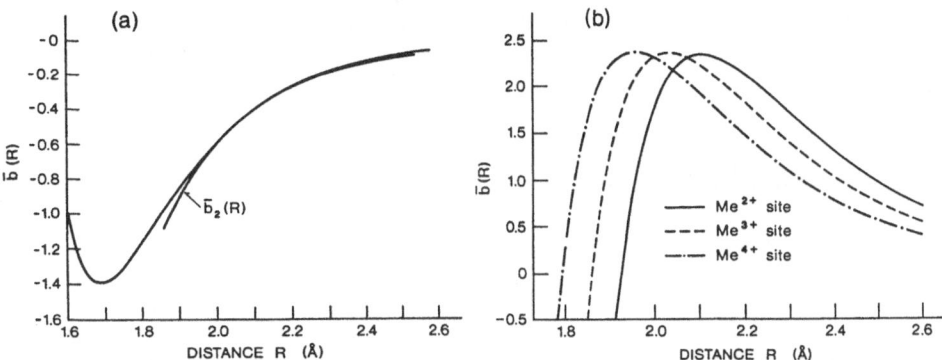

Fig. 1.2. (a) $b(R)$ for Fe^{3+} in MgO (on Me^{2+} sites) with the $\bar{b}_2(R)$ power-law dependence $\bar{b}_2 = -0.41(2.101/R)^8$; (b) $\bar{b}(R)$ for Cr^{3+} on Me^{4+}, Me^{3+} and Me^{2+} sites in octahedral oxygen coordination. From [1.14]

be extended to d ions with half-filled e_g or t_{2g} subshells. This has recently been done for Cr^{3+} $S = \frac{3}{2}$ $L' = 0$ in oxygen octahedral coordination by *Müller* and *Berlinger* (MB) [1.14], the Cr^{3+} replacing host ions with valencies two, three and four. The intrinsic $\bar{b}_2'(R)$ parameters computed with (1.10) are shown in Fig. 1.2b. Owing to the electronic $(t_{2g})^3$ configuration of Cr^{3+}, $\bar{b}_2(R)$ is of opposite sign to that of Fe^{3+} with configuration $(t_{2g})^3 (e_g)^2$. For Fe^{3+}, $\bar{b}_2(R)$ is negative and has a minimum at $\simeq 1.7\,\text{Å}$, an atomic separation not occurring in crystals. On the other hand, $\bar{b}_2(R)$ of Cr^{3+} has a maximum at $\simeq 2.0\,\text{Å}$ for a Me^{3+} replacement. From (1.8), one therefore infers that this ion is mainly sensitive to the angles Θ and ψ as is Gd^{3+} with parameter t_2 near zero [1.10]. In contrast, for Fe^{3+} with a $t_2 = 8 \pm 1$ dependence the b_2^m are very sensitive to R in that range. Replacement of Fe^{3+} and Cr^{3+} on the same lattice site is therefore very helpful in the analysis. The quantitative applicability of the proposed $\bar{b}_2(R)$ dependence in the trigonal $LaAlO_3$ and Al_2O_3 crystals has been demonstrated [1.14]. Figure 1.2b also shows a charge-misfit dependence on the host ions of Cr^{3+}. This is a further limitation of the model not considered originally by *Newman* and *Urban* [1.10] but predicted theoretically by *Sangster* [1.39].

In an octahedral environment, the Cr^{3+} ground state is $^4A_2$. Thus, in a superposition model for ionic systems the expressions on the right-hand side of (1.8) each have to be replaced by a sum of two terms [1.40]. In this model, only ^4T excited states were taken into account, and ^2E states and covalency were neglected. However, these are important. An application of the ionic model to $Cr^{3+}:Al_2O_3$ yielded a ground-state splitting b_2^0 four times smaller than that observed [1.41], in contrast to the result of using $\bar{b}_2(R)$ of Fig. 1.2 and (1.8). The excited ^2E states for Cr^{3+} and the quartets for Fe^{3+} have comparable excitation energies, of the order $2\,\text{eV}$ or even less. This makes plausible why the accepted applicability of the single parameter $\bar{b}_2(R)$ superposition model for Fe^{3+} also entails that for Cr^{3+}. Recently, *Yeung* and *Newman* have analyzed the superposition model for Cr^{3+} in detail, taking into account the above-mentioned excited states [1.42]. The success of the MB model has been explained as a consequence of the stability of the ratio $\bar{B}_4/\bar{B}_2$, together with the acceptance of small negative values of the parameters $\bar{t}_2$ as being realistic. $\bar{B}_4$

and $\bar{B}_2$ are the known crystal field parameters. In Yeung and Newman's reinterpretation of the experimental data, parameter values are obtained which are reasonably consistent with optical values.

1.3 Experimental Techniques

1.3.1 Standard Equipment

We attempt here to give a brief outline of how the resonance principle, described at the beginning of this section, can be experimentally realized. Detailed experimental information is contained in the book by *Poole* [1.43]. An elaborate experimental design is helpful to improve the sensitivity of the method and to save time. The EPR spectrometer consists essentially of a microwave emitter, usually a Klystron, a resonance device containing the sample and a receiver. The resonance device – commonly a resonant cavity – is placed between the poles of a magnet, yielding a homogeneous horizontal field. This magnet is mounted on a vertical axis, allowing rotation of the magnetic-field direction. Typical microwave frequencies and magnetic fields (for a g factor of 2, free electrons) range from 3.2 GHz/0.114 Tesla (S-band) to 35 GHz/1.25 Tesla (Q-band). Since typical EPR linewidths are of the order of a few Gauss (10^{-4} Tesla), the external magnetic field must be stabilized to less than 10^{-5}. We refer to the literature [1.43] for a review of the different spectrometer concepts. Here, we shall concentrate on specific aspects of the resonant cavity. From time-dependent perturbation theory, it follows [1.3] that the probability for an electron spin to make a transition from the lower to the upper Zeeman state, induced by the high-frequency field H_1, is given (for $g = 2$) by

$$W = \left(4\mu_B^2 H_1^2/\hbar^2\right) f(\omega)\langle S, M | S_\pm | S, M - 1\rangle^2 \tag{1.11}$$

where $f(\omega)$ is the line-shape function. The transition probability is the same for upward and downward transitions. The number of upward and downward transitions in time dt is then given by [1.11]

$$dN_\uparrow = W \cdot N_1 \, dt \quad ; \qquad dN_\downarrow = W \cdot N_2 \, dt \tag{1.12}$$

where N_1, N_2 denote the spin populations of lower and upper levels, respectively. Each upward transition absorbs an energy quantum of $\hbar\omega_{\mathrm{res}}$ from the microwave radiation, while each downward transition generates one. Therefore, the total energy absorption per unit of time is

$$\frac{dE}{dt} = W\left(N_1 - N_2\right)\hbar\omega_{\mathrm{res}} \tag{1.13}$$

where ω_{res} is the resonant microwave frequency (Larmor frequency). As long as relaxation is fast enough to keep N_1 and N_2 at thermal equilibrium, (1.11) and (1.13) show that the rate of energy absorption is proportional to H_1^2. For an optimal signal-to-noise ratio, the experimental set-up must therefore fulfill the following requirements:

- The microwave magnetic field H_1 should be perpendicular to the Zeeman field H_0, owing to the resonance condition for allowed transitions;
- H_1 should be maximum at the sample site to ensure the maximum rate of energy absorption;
- the sample should be placed at a position of minimum microwave electric field E_1 to keep dielectric losses as low as possible.

These aimes are commonly achieved by the use of a resonant cavity, although other devices such as helices are sometimes employed. In the cavity, the microwave builds up a standing wave pattern. Figure 1.3a depicts a cylindrical cavity with the TE_{011} mode together with a sample placed at highest H_1 and lowest E_1 field intensities. It also shows a variable microwave coupling arrangement which allows the rf field intensity to be varied within the cavity. This is necessary in order to tune the spectrometer to highest sensitivity and, in special cases (low relaxation rate), to avoid saturation ($N_1 = N_2$) of the spin resonance.

The cavity is characterized by its quality factor Q expressing the ratio between its energy content and its energy loss per period [1.43]. At resonance, the effective Q factor of the cavity/sample system is lowered, since energy is absorbed by the sample and dissipated to the lattice by relaxation. The sensitivity is best when the Q factor of the unloaded cavity is as high as possible [1.43]. In favorable cases, the minimum detectable number of spins lies in the range 10^{11} to 10^{12} at room temperature (assuming a linewidth of 1 G) [1.4]. Samples with elevated high-frequency dielectric constant or a loss can lower the Q value and hence the sensitivity. This may be crucial for the investigation of ferroelectric phase transitions, where the increased dielectric constant detunes the cavity and the loss shoots up. When the dielectric constant does not increase significantly with decreasing temperature, the sensitivity is improved at lower temperatures. Good signals are usually obtained from 10^{16} to 10^{17} spins. With samples of about $0.1 \, cm^3$, this corresponds to concentrations of magnetic impurities of 10 to 100 ppm. Often, crystals need not even be intentionally doped to achieve this concentration of resonance centers. Many natural minerals and commercially "pure" crystals show concentrations in this range of, for instance, Fe^{3+} or Mn^{2+}. To improve the sensitivity further, a lock-in technique is applied for signal detection [1.43], commonly by modulating the Zeeman field H and therefore the absorption amplitude. As a consequence, the derivative of the signal as a function of magnetic field is recorded. Resonance lines are then displayed on the $x - y$ plotter as first derivatives. Very weak resonances can be detected by the application of a signal-averaging technique using a multichannel analyzer.

It should be noted that an EPR "spectrum" taken with a resonant cavity does not represent a frequency scan. Since the cavity Q is high for a very sharply defined frequency only, it is not possible to change the rf frequency over a wide range. Instead, the spectrum is obtained by scanning the external magnetic field H, and thus changing the Zeeman splitting of the electronic energy levels. The EPR spectrum therefore consists of resonances as a function of field intensity $|H|$ as described in Sect. 1.2.2 (Fig. 1.1).

1.3.2 Special Requirements for Investigations
of Structural Phase Transitions

The investigation of phase transitions by EPR demands the ability to apply various external variables to the sample. The most important variable is the temperature. A flexible instrumentation should be able to cover a range between a few Kelvin and some 1000 K. Very accurate temperature stabilization at a certain nominal value is not only important for detailed, reproducible investigations of the temperature dependence of the transition, but, as in the case of $SrTiO_3$ with its 105 K transition, may be crucial owing to the strong temperature dependence of the dielectric constant. This dependence changes the capacitive component of the cavity, and hence detunes it. Temperature stabilization should keep the nominal value within a range of 0.1 to 0.001 K, depending on the sample and transition characteristics. Reference [1.44] describes a versatile low-cost version of the commonly used reflection cavity. The cavity is water-tight and is either directly immersed in liquid helium or brought into thermal contact with it by a cold finger attached to the base (Fig. 1.3A). The cavity, made from brass, thus constitutes an almost isothermal enclosure for the sample centered therein. Short-time temperature variations are further damped at the sample site since the cavity walls are not in direct heat contact with the sample. For temperatures above 77 K, the cavity is in the empty helium dewar immersed in liquid nitrogen. Intermediate temperatures between 4.2 K and room temperature are obtained by heating the cavity walls with a proportionally controlled heater wound around the outer wall [1.44]. The temperature is sensed by a thermocouple inside the cavity walls. The short-time fluctuations observed in the wall at 100 K do not exceed approximately ± 0.025 K, and at $T < 40$ K these variations are even smaller. Therefore, the temperature stability at the sample in the center is around ± 0.01 K. An improved version of this cavity setup [1.45] has even reached stabilities of 0.7 mK at the sample site. Such a stability of $\Delta T/T \leq 10^{-5}$ makes critical SPT investigations of this type competitive with the best ones on magnetic phase transitions.

Other variables during investigations on phase transitions are pressure and uniaxial stress. For more specialized cases, it may also be necessary to irradiate the sample with light or to apply a static electric field. These treatments should be applied consecutively or even simultaneously, and without the necessity of heating the sample up to room temperature. Such flexibility of the instrumentation saves much time, especially when working at very low temperatures. In Ref. [1.44] and other references cited therein possible solutions are presented to these requirements using a modular form. For such purposes, a cylindrical TE_{011} cavity instead of a rectangular one is especially helpful, although the latter is smaller, since the cylindrical cavity allows a central access from outside. Different experimental arrangements for various conditions (Fig. 1.3B) have been realized by mounting the sample on a rod introduced through a tube following the wave guide. This necessitates exact orientation and grinding of the sample prior to mounting. To avoid this, the sample can also be glued onto a small goniometer attached inside the cavity base [1.17], but this arrangement does not allow access to the sample from outside. The modular form also permits simultaneous detection of nuclear and electron resonance, called electron nuclear double resonance or ENDOR [1.3, 4]. This technique makes it pos-

(A)

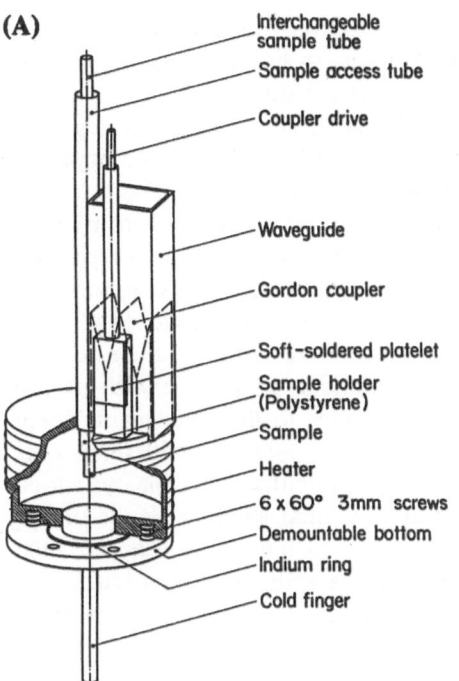

Interchangeable sample tube
Sample access tube
Coupler drive
Waveguide
Gordon coupler
Soft-soldered platelet
Sample holder (Polystyrene)
Sample
Heater
6 x 60° 3mm screws
Demountable bottom
Indium ring
Cold finger

Fig. 1.3. (A) Low-temperature cavity with central sample access tube and eccentric microwave variable coupling arrangement. From [1.44]. (B) Semi-schematic cross section of the modular sample arrangements used in the LT cavity. The different mountings show: (a) sample with known plane; (b) sample in known direction; (c) and (d) powder samples; (e) irradiation with light, $h\nu$; (f) application of electric field $E \perp H_{rf}$; (g) ENDOR experiment; (h) uniaxial compression; p; (i) uniaxial compression, p, under light excitation, $h\nu$; (k) electric field $E \| H_{rf}$; and (l) tension experiment. From [1.44]

(B)

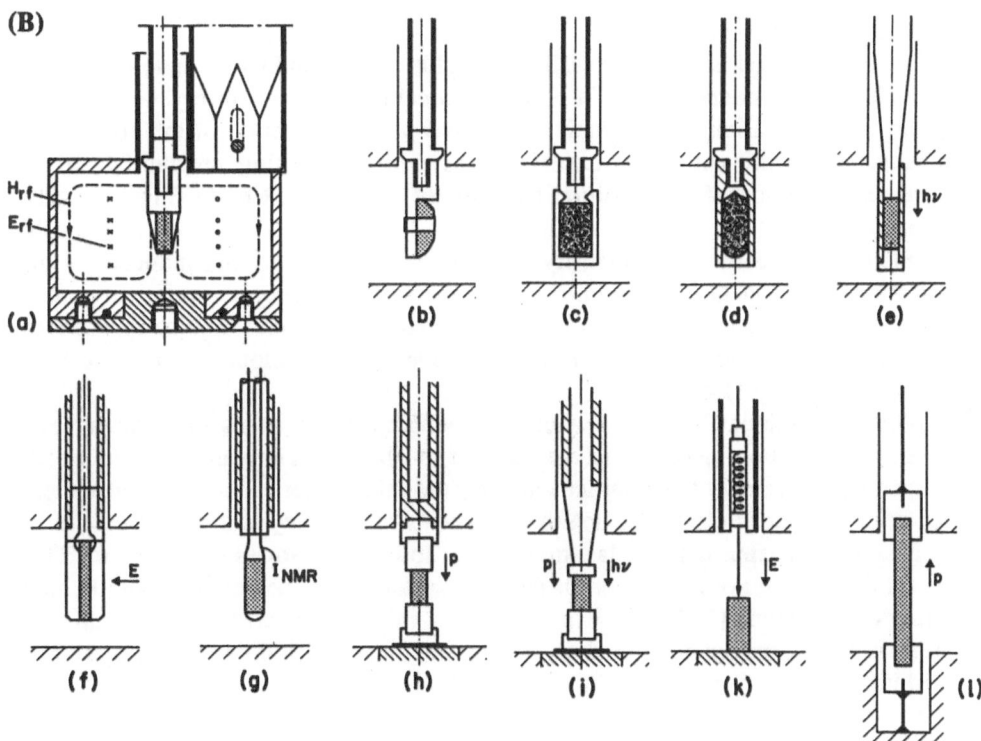

sible to determine hyperfine constants and nuclear g factors with very high precision [1.4]. This in turn enables high-order hyperfine effects to be investigated, yielding detailed information on hyperfine interactions with ligand ions.

The cavity described above allows one to investigate SPT's at temperatures down to the pumped helium range in which quantum effects of SPT's become apparent. However, in oxides, SPT's also occur well above room temperature. The transition in $LaAlO_3$ with $T_c \simeq 797$ K is the most prominent example so far studied in detail, including its behavior under uniaxial stress. A cavity used to study the temperature dependence accurately has been described earlier [1.44]. Another cavity, in which uniaxial stress can be applied, has recently been developed by *Berlinger* [1.45], and appears to be the only one in use for such high-temperature work. Its main features are shown in Fig. 1.4a. The cavity is again of cylindrical TE_{011} type with its axis aligned along the stress to be applied and perpendicular to the constant magnetic field. The wall of the cavity is kept at room temperature by cooling it with three loops of a water-conducting brass pipe soldered to its outside circular wall. The cylindrical sample is aligned along its center axis, and compressional stress is applied along this axis via two quartz spacers (Fig. 1.4b). Heating is achieved by a quartz tube, into which the sample and quartz cylinders fit and onto which platinum wires have been applied along the outside of the tube. The wires are well separated so that the microwaves can reach the sample through the tube. A regulated dc current generates Joule heat. To homogenize the temperature, quartz wool is wrapped around the quartz heater tube. Recent improvements of the regulation system ensure a temperature stability of $\Delta T \simeq 10$ mK at 1300 K, i.e., a relative stability of $\Delta T_T \simeq 7.7 \times 10^{-6}$, comparable to that achieved in the low-temperature cavity [1.45].

The EPR method is generally easily applicable to nonconducting materials doped with suitable paramagnetic ions. It is not practicable, however, in crystals containing large concentrations of magnetic ions; the dipolar interaction in the absence of exchange narrowing can cause excessive linewidth broadening precluding an accurate determination of line position and thus of symmetry. Conducting materials, e.g. metals, lower the Q value of the cavity when placed at sites of nonvanishing microwave electric-field intensity and owing to eddy currents induced by the magnetic rf component. Furthermore, due to the skin effect, only a relatively small portion of the sample bulk may be probed. For several special applications of the EPR method, other experimental principles have been developed which are now summarized briefly [1.46].

With conventional Zeeman modulation, certain charge states of paramagnetic ions of the transition-metal group are difficult to observe. They are those in which the ion couples strongly to the lattice either by the Jahn-Teller effect or because there is an even number of electrons such that Kramer's degeneracy is absent. Because of this coupling, the spin-lattice relaxation time at low temperatures can be short. This enables the microwave absorption in the spin system to be detected by the thermal heating of the lattice [1.47, 48]. Experimentally, this is realized by a small carbon-film resistance thermometer, which is in thermal contact with the sample [1.49]. For broad lines and short relaxation times, this technique can be significantly superior to the conventional one because the microwave power can then be increased without saturating the level occupancies.

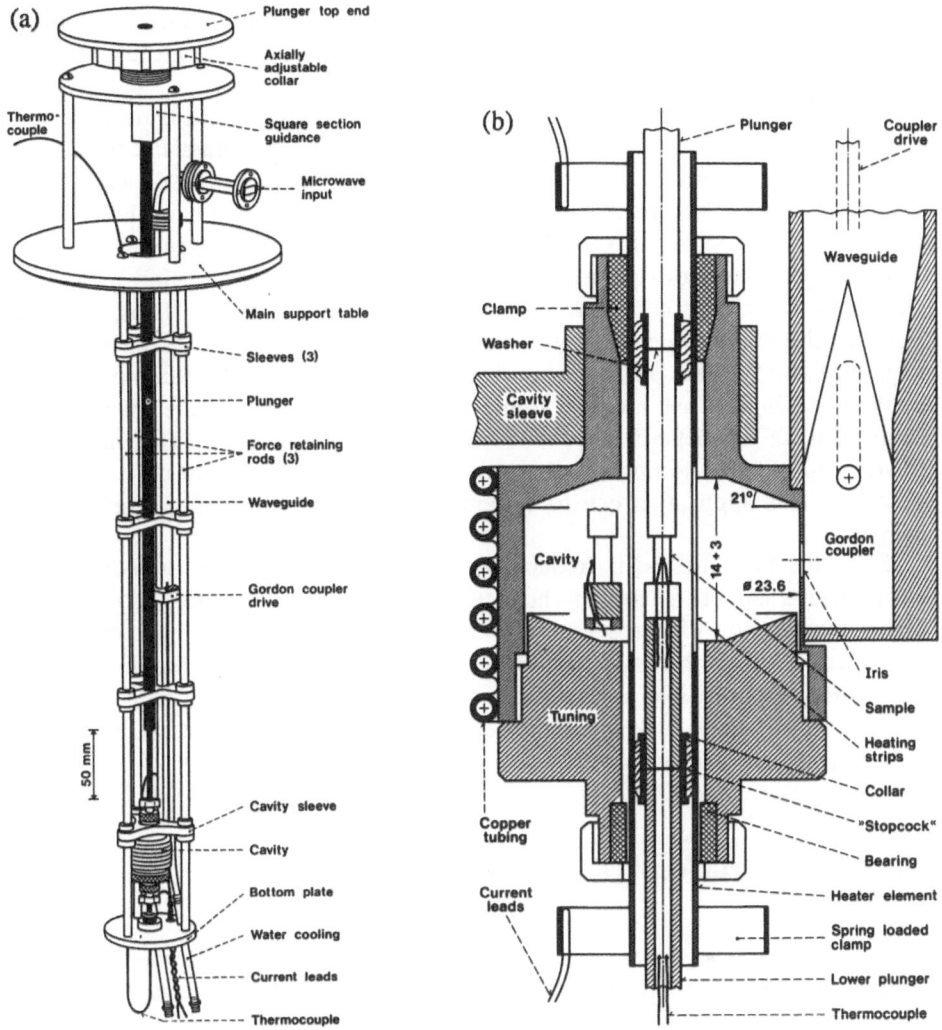

Fig. 1.4a,b. High-temperature cavity for EPR in presence of uniaxial stress. From [1.45]

While for Kramer's ions the coupling to electric fields is generally small [1.3, 26], this is not the case for paramagnetic ions with an even number of spins or for Kramer's ions occupying sites which lack inversion symmetry [1.50, 51]. For these cases, modulation of an applied electric field allows an accurate determination of the strength of the electric interaction. From the shapes of the lines, the distribution of the local electric field can be found [1.52].

Coupling to elastic strain is another possibility for detecting electron paramagnetic resonances. In this technique, the strains are modulated by an external transducer attached to the sample [1.53, 54], operating in the range 20 to 60 kHz. In addition, a magnetic-field modulation is also applied and the ESR signal is detected at both modulation frequencies. The main advantage of this method appears to lie in the observation of broad lines which result from coupling to inhomogeneous elastic-strain fields in the crystal due to large coupling constants, as in part also observed near SPT's.

By means of a variable frequency technique, it is possible to measure the crystal-field splittings of paramagnetic centers at fixed or even zero magnetic field [1.10]. This can be very useful to confirm or even obtain the parameters of the Hamiltonian in difficult cases. The variable frequency system uses a broad-band helix instead of a cavity [1.10]. The helix is a slow wave structure and replaces the cavity. It is advantageous for low-temperature experiments because it is small and can be operated with smaller microwave power densities than a cavity. Thus, saturation problems should not occur. Although the effective Q factor of the helix is orders of magnitude smaller than that of a cavity, this disadvantage may be compensated by broad-band coupling and a higher filling factor, which expresses the fraction of the cavity or helix magnetic-field volume filled by the sample [1.3]. This method has proved helpful for observing signals in materials with conductive or dielectric losses ($KNbO_3$). This application can be important in photoconductors, ferroelectrics or metals.

Since the EPR method is frequently applied in the chemistry of radicals, EPR spectrometers are often used in chemical laboratories. Application of the method to the investigation of phase transformations is thus possible without much extra investment. Such attributes as sensitivity, relative simplicity, precision in measuring local symmetries, flexible sample treatment and variable magnetic field H both in intensity and direction, make the EPR method extraordinarily helpful and informative for the investigation of structural phase transitions.

1.4 Oxide Ferroelectrics

Ferroelectrics had long been investigated before EPR was ever used, the main reason being their accessibility via dielectric measurements. With these, the static properties of interest, namely the order parameter and the susceptibility could be determined. These two quantities are proportional to the polarization $P(T)$ and the dielectric constant $\varepsilon(t) \gg 1$. Furthermore, since large displacements occur, they are also accessible to conventional X-ray techniques. Together with birefringence measurements, the symmetry properties of the lower-symmetry phases were obtained, the piezoelastic and electrostrictive properties being known. We nevertheless begin our review with this field in order to retain the sequence followed by the articles in Vol. I of this series. Furthermore, it allows us to introduce the method with reference to a known quantity $P(T)$, the order parameter of the ferroelectric (Sect. 1.4.2). Three sections, 1.4.2–4, are devoted to recent, quite successful applications of the super-position model introduced in Sect. 1.2.4. The sequence of these allows the reader to appreciate the scholarly application and progress of EPR in the field. It will be shown that the recent EPR results point to the microscopic origin of the oxide ferroelectrics. Furthermore EPR supports a substantial order-disorder character of their dynamics. In addition, magnetic resonance should prove most helpful in determining $P(T)$ in the quantum low-temperature range, where locked surface charges make dielectric hysteresis-loop measurements tedious if not impossible.

1.4.1 EPR of Fe^{3+} and Gd^{3+} in Cubic and Tetragonal $BaTiO_3$

The most thoroughly investigated ferroelectric crystal in this category is $BaTiO_3$. It belongs to the well-known perovskite family, where each Ti ion is surrounded by an oxygen octahedron, whereas the Ba is twelve-fold coordinated [1.55], see Fig. 1.5. The ferroelectric phase transition essentially results from a cooperative motion of the Ti sublattice against the highly polarizable oxygens. Fe^{3+} EPR spectra of unintentionally, and later intentionally, Fe-doped Remeika (KF flux) crystals were investigated and their temperature dependence measured. The first complete study characteristic for this type of investigation was carried out by *Hornig* et al. [1.56], where five allowed $\Delta M = 1$ lines were observed. In the high-temperature cubic phase, the Hamiltonian of the $S = \frac{5}{2}$ spectrum with the single constant a is as given by (1.6) and a cubic spectrum was observed. From the ionic radius of the Fe^{3+} of 0.64 Å, close to that of the Ti^{4+} of 0.63 Å, it was concluded that Fe^{3+} substitutes for Ti^{4+} on the octahedral site.

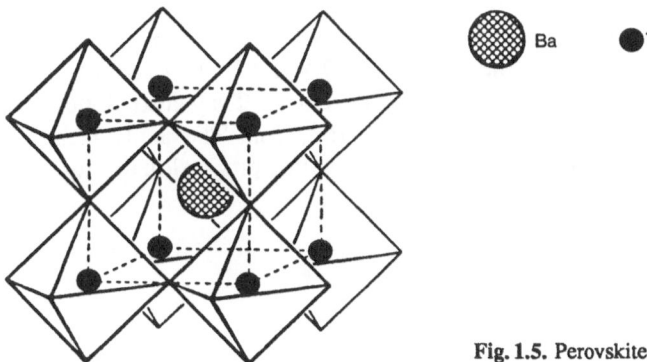

Fig. 1.5. Perovskite structure of $SrTiO_3$. From [1.15]

Below the cubic ($Pm3m$) to tetragonal ($P4mm$) ferroelectric transition at 120° C, an axial term of the form

$$b_2^0 O_2^0 \equiv D \left[S_z^2 - \tfrac{1}{3} S(S+1) \right] \quad , \tag{1.14}$$

had to be added to account for the data, with z along the [001] tetragonal domain axes. Such crystals are called c-domain crystals. The (001) platelets were electrically polarized parallel to the [001] direction. Thus only one spectrum was observed, since (1.14) does not distinguish between positive and negative polarization directions z. The quantity D was found to be of the order of the X-band spectrometer frequency, and necessitated a diagonalization of the full Hamiltonian. Figure 1.6 displays the temperature dependence of the quantity $\Delta = D(T) + a(T)$. From it, one clearly sees the discontinuity in $D(T)$ which drops to zero at the first-order transition. There, only the temperature-independent cubic splitting a persists. The figure makes it

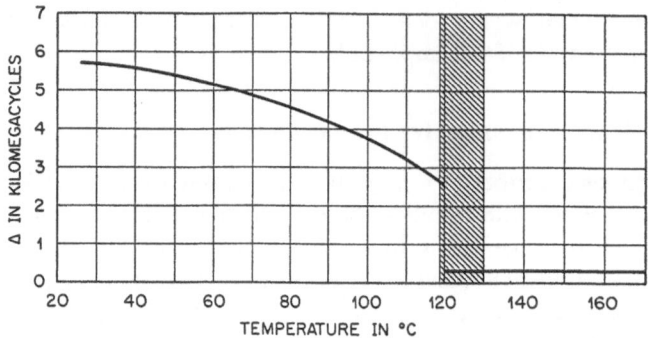

Fig. 1.6. Temperature dependence of the parameter Δ of Fe^{3+} in $BaTiO_3$. Reprinted with·permission from [1.56]. Copyright 1959 Pergamon Journals Ltd.

evident that $D(T)$ is related to the order parameter of the ferroelectric transition and the substitutional Fe^{3+} detects coherent displacements of intrinsic ions. $D = b_2^0 = \frac{1}{3}B_2^0$ has to vary linearly with the crystal-field potential $V_2^0 = \partial^2 V/\partial_z^2$ existing only in the tetragonal phase. The latter depends linearly on the piezoelectric strain of the crystal $\delta a/a$, because the tetragonal phase results from the cubic structure containing a center of inversion $\delta a/a \propto P^2(T)$. Thus one expects $D(T)$ to vary in proportion to the square of the polarization. This was verified by *Rimai* and *De Mars* [1.57] as shown in Fig. 1.7; they obtained $D = 1.6P^2$ Oerstedts if P is measured in $\mu Coulomb/cm^2$ and D in units of $10^{-4} cm^{-1}$. They also doped the crystal intentionally with Gd^{3+} $\left(S = \frac{7}{2}\right)$ replacing a Ba^{2+} ion. The analysis is the same as just described for Fe^{3+}. The b_2^0 parameter also varies proportionally to $\delta a/a$ and $P^2(T)$, as shown.

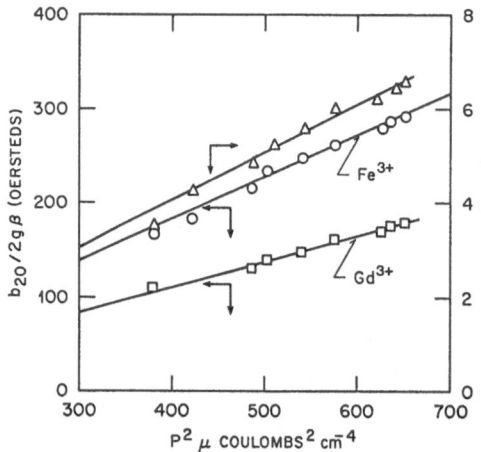

Fig. 1.7. Dependence of the polarization of the piezoelectric strain and the second-order spin-Hamiltonian parameter of Fe^{3+} and Gd^{3+} in tetragonal $BaTiO_3$. From [1.57]

Wemple [1.58] calculated the axial parameter in the presence of a lattice polarization assuming it results solely from the polarization-induced lattice strain $\varepsilon \simeq 10^{-3}P^2$. Then, using the measured proportionality to strain of $D = 0.14 \times \varepsilon$ [cm^{-1}] for Fe^{3+} in MgO, he obtained $D = 1.4P^2$. He also proved that this holds to within a factor of 2 for electric-field induced polarizations at 4.2 K in cubic $KTaO_3$. As we shall see in the next section, it was of importance that he used the values for MgO where the Fe^{3+} is *centered*. Later, the relation $D = 1.6P^2$ for Fe^{3+} was also found to hold in tetragonal ferroelectric $PbTiO_3$ [1.59].

From these early studies, one can draw the following conclusions:

a) The large splittings observed make EPR a sensitive probe of the occurrence of structural changes, and one can also determine whether they are of first or second order.

b) With EPR, investigations of the change in the order parameter as a function of temperature are possible. At low temperatures, this is of importance for ferroelectrics because the surface charges become so immobile that dielectric measurements of the hysteresis loop can be hindered [1.60].

c) From EPR, the orientation and relative number of domains can readily be measured since the paramagnetic impurities are statistically distributed among them.

d) From the O_n^m symmetry axes, the *local* point symmetry can be determined.

1.4.2 Fe^{3+} and Gd^{3+} Spectra in Orthorhombic and Rhombohedral $BaTiO_3$

Fe^{3+} in $BaTiO_3$ has subsequently been studied in the orthorhombic $Cmm2$ and rhombohedral $R3m$ phases by *Sakudo* [1.61] and by *Sakudo* and *Unoki* [1.62]. In these phases six and eight domains occur, respectively; thus the spectra overlap so much that they are difficult to analyze. The Japanese authors therefore biased their samples with electric fields along the [110] and [111] cubic axes. In doing so, they obtained samples with domains oriented parallel to the applied biasing fields. Their results are summarized in Table 1.1 (the g value was constant; $g = 2.0036$). From this table, a number of remarkable facts are seen: (i) the cubic-splitting constant $a \simeq 0.01\,\mathrm{cm}^{-1}$ does not vary by more than 10% across the phase changes. This indicates that the Fe^{3+} always sees the same octahedral oxygen environment; (ii) most important, the axial $D \equiv b_2^0$ parameter changes sign on going from the tetrahedral to the orthorhombic phase and no orthorhombic term $E(S_x^2 - S_y^3) = b_2^2 O_2^2$ appears. The z-axis of the O_2^0 term is parallel to [100] and lies perpendicular to the polarization axes [110]; and (iii) the axial D term in the rhombohedral phase is parallel to the polariaztion ($P\|[111]$), but its size is one to two orders of magnitude smaller than in the tetragonal and orthorhombic phases, although the polarization is of the same order in all polar phases [1.55].

Table 1.1. EPR parameters of Fe^{3+} in $BaTiO_3$

Crystal structure	Temp.	Ref.	a [cm^{-1}]	b_2^0 [cm^{-1}]	z-axis direction	Ref.	Centered Fe^{3+} computed b_2^0 [cm^{-1}]
Cubic	160°C	[1.56]	+0.0102	0			
Tetragonal	27°C	[1.56]	+0.0091	+0.0929	$\langle 100 \rangle \| P$	[1.63]	$+0.105 \pm 0.025$
Orthorhombic	−60°C	[1.61]	+0.0094	−0.0053	$\langle 001 \rangle \perp P$	[1.63]	-0.057 ± 0.014
Rhombohedral	−196°C	[1.62]	+0.0115	−0.0023	$\langle 111 \rangle \| P$	[1.63]	-0.0083 ± 0.002

Sakudo [1.61] discussed his results in terms of a model of consecutive ordering of *oxygen* polarizations along [100] (tetragonal phase), [100] and [010] (orthorhombic), and finally [100], [010] and [001] (rhombohedral) directions. Later *Takeda* reported on his Gd^{3+} studies in the lower two phases of $BaTiO_3$ [1.64]. He found, in contrast to the Fe^{3+} results, axial $|b_2^0|$ terms, directed along $\langle 110 \rangle$ and $\langle 111 \rangle$ axes, of almost the same magnitude of 267 and $340 \times 10^{-4}\,cm^{-1}$, respectively, in the tetragonal phase. Thus, they are perfectly normal in their direction and magnitude to the observed polarizations in the three phases. *Takeda* computed the field gradient for Gd^{3+} resulting from the neighboring oxygen positions including their polarization, as *Sakudo* did for Fe^{3+}, the contributions from the Ti^{4+} being small. He obtained reasonable agreement on assuming oxygen polarizations parallel to $\langle 110 \rangle$ and $\langle 111 \rangle$, but disagreement with the Sakudo model of polarizations parallel to $\langle 100 \rangle$.

Actually, *Sakudo* [1.61] had expressed some doubt about this interpretation, because, in the NMR results for $KNbO_3$, an orthorhombic nuclear quadrupole splitting of the Nb nucleus was observed in the orthorhombic phase [1.65]. However, what was not sufficiently known at the time was the sensitivity difference of EPR and NMR to the inner electric field and covalency [1.66]. The EPR b_2^0 parameter in oxides is mainly determined by the next-neighbor *positions*, whereas the nuclear quadrupole splitting is proportional to a slowly converging function of summations of the individual monopole, dipole and even quadrupole contributions [1.67]. Thus, NMR and EPR do not measure the same property even if the probing Fe^{3+} ion were located exactly at the Ti^{4+} site. In the following, the successful use of the superposition model to calculate the b_2^m parameters with good numerical precision is summarized.

Siegel and *Müller* [1.63] used this model for Fe^{3+} in $BaTiO_3$ to account for the Fe^{3+} b_2^m data of Table 1.1. It turned out that the Fe^{3+} participates in the collective off-center Ti^{4+} displacement by no more than 10%, i.e., it remains approximately in the center of the octahedron. In the last column of Table 1.1, computed values of b_2^0 are listed using (1.8) *without adjustable parameters*. They were obtained with the intrinsic constants $\bar{b}_2(R_0)$ and $t_2 = 8 \pm 1$ from Fig. 1.2a, assuming a centered Fe^{3+} in the octahedron and using for the oxygens their intrinsic positions as determined from refined X-ray scattering analysis. Table 1.2 compiles the known experimental Fe^{3+} EPR parameters of $PbTiO_3$ in the tetragonal and $KNbO_3$ in the orthorhombic phase. Note that with this analysis a near axial term with $Z \perp P$ in the orthorhombic phase, i.e., $b_2^2 \ll b_2^0$, was obtained. In the rhombohedral $BaTiO_3$ phase, the oxygens are located on an almost undistorted octahedron. As the Fe^{3+} is at its center, the axial b_2^0 term almost vanishes. The success of this analysis implies that the polarization

Table 1.2. EPR parameters of Fe^{3+} in $PbTiO_3$ and $KNbO_3$

Crystal	Structure	Temp.	Ref.	b_2^0 [cm^{-1}]	z-axis direction	Ref.	Centered Fe^{3+} computed b_2^0 [cm^{-1}]
$PbTiO_3$	tetragonal	300 K	[1.59]	$+0.53 \pm 0.02$	$\langle 100 \rangle \| P$	[1.63]	$+0.58 \pm 0.14$
$KNbO_3$	orthorhombic	300 K	[1.68]	$|+0.18 \pm 0.001|$	$\langle 001 \rangle \perp P$	[1.63]	-0.15 ± 0.04
				$|b_2^2/D| = 0.17$		[1.63]	$|b_2^2/D| = 0.12 \pm 0.1$

fields can be neglected as compared to the nearest-neighbor positions, owing to ionic and covalent effects. It justifies the assumption of *Wemple* [1.58] in deriving the $D = 1.4P^2$ formula from polarization-induced strain only, neglecting the direct polarization effect on the centered Fe^{3+} ion.

The reduced participation of the Fe^{3+} in the collective Ti^{4+} motion implies that on doping, the transition temperature T_c of $BaTiO_3$ must be strongly depressed. This property has indeed been found: a doping of 1% reduced T_c by 20°C [1.69]. The Fe^{3+} contains five $3d$ electrons ($3d^5$) in antibonding or nonbonding orbitals. They impede bonding of the oxygen p electrons with the empty $3d$ orbitals, as is possible in Ti^{4+} ($3d^0$) or Nb^{4+} ($4d^0$). *This points to the very origin of ferroelectricity in oxides containing ions with empty d orbitals.* Of course, the size and charge of the ion are also important. With regard to size, Zr^{4+} has empty $4d$ orbitals, configuration $4d^0$, like Nb^{5+}, but suppresses T_c upon doping into $BaTiO_3$, because its size of 0.79 Å is 0.11 Å larger than that of Ti^{4+} or Nb^{5+}. However, it depresses T_c less than Sn^{4+} with a radius of only 0.71 Å but a full $4d^{10}$ shell [1.55]. This shows that for Zr^{4+} the empty $4d$ shell, which favors ferroelectricity, counteracts the suppressive effect of the large size. However, coming back to Fe^{3+}, its charge misfit also reduces its participation in the cooperative motion. Indeed its effective charge with respect to the lattice is -1. Thus the surrounding oxygens are repelled owing to this effect, as also shown by calculations of *Sangster* [1.39]. This charge-misfit effect adds to that of the half-filled $3d$ shell. As will be discussed in the next section, the charge misfit also causes the Cr^{3+} in $BaTiO_3$ to remain but barely centered despite the absence of the two antibonding $(e_g)^2$ electrons, which are present in Fe^{3+}. Finally, Mn^{4+} with no charge misfit but the same electron configuration as Cr^{3+} indeed follows the Ti^{4+} motion.

1.4.3 Cr^{3+} in $BaTiO_3$: A Centered Ion with a Flat Local Potential

There are two possible reasons why the Fe^{3+} at the Ti^{4+} site should be centered: the charge misfit by minus one unit and the presence of two (e_g) antibonding orbitals. To further elucidate the situation, it was decided to take another trivalent ion with nearly the same ionic radius as Fe^{3+} but with a different $3d$ configuration. Cr^{3+} is such a ion, with a $3d^3$ electron configuration compared to the $3d^5$ of Fe^{3+}. Whereas Fe^{3+} in the high-spin configuration has its two subshells with t_g and e_g character half-filled, e.g. $(t_{2g})^3$, $(e_g)^2$, Cr^{3+} has only the t_{2g} subshells half-filled with $(t_{2g})^3$ and the e_g shell empty. The $(t_{2g})^3$ are essentially nonbonding with their charge density pointing midway between the repelling oxygen electron density.

To quantitatively investigate the Cr^{3+} position in $BaTiO_3$, a paramagnetic resonance study of Cr^{3+} was carried out after doping a single crystal [1.70]. EPR spectra were investigated in all four phases as a function of temperature. The spectra obtained were very different from those observed for Fe^{3+} as far as the size of the splittings, their sign and the orientation of the main magnetic axis in the orthorhombic phase are concerned. In each ferroelectric phase (FEP), the largest splitting occurred along the ferroelectric domain axis, i.e., a term of the form (1.14) with z along the $\langle 100 \rangle$, $\langle 110 \rangle$ or $\langle 111 \rangle$ axis in the tetragonal, orthorhombic and rhombohedral phases, respectively. In the orthorhombic phase, a nonvanishing $E(S_x^2 - S_y^2)$ fine-structure

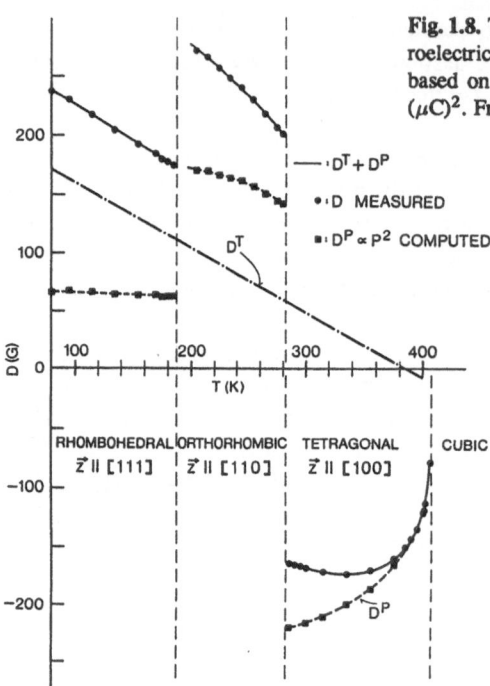

Fig. 1.8. Temperature dependence of $D(T)$ in the three ferroelectric phases of BaTiO$_3$. The theoretical analysis is based on (1.15) with $\beta = 0.56\,G/K$ and $\alpha = 0.28\,G\,cm^4/(\mu C)^2$. From [1.70]

splitting was also measured. The analysis of the main EPR crystal-field terms $D(T)$ in the Hamiltonian for the three FEP's between 100 K and the highest T_c could be accounted for by just two terms for each phase, namely a term D^P proportional to the square of the lattice polarization $P(T)$, and a large term, D^T, linear in temperature, with the latter being the same for all FEP's:

$$D(T) = D^P + D^T = \alpha_i P_i(T)^2 + \beta (T - T_0) \qquad (1.15)$$

with $i =$ t, o, r denoting the tetragonal, orthorhombic and trigonal ferroelectric phases, respectively, and $\beta = 0.56\,G/K$. Figure 1.8 reproduces the EPR $D(T)$ values in the three phases with the analysis according to (1.15) and $\alpha_t = 0.29\,G\,cm^4/(\mu C)^2$ assumed the same for all phases.

The term proportional to $P^2(T)$ is direct evidence that the Cr^{3+} remains centered. However this is only barely so because the term linear in T, not observed for Fe^{3+}, is a sign of large local Cr^{3+} fluctuations. These are confirmed by the low-temperature behavior of $D(T)$ between 4.2 and 100 K, which can be accounted for by an Einstein model for the Cr^{3+} with quite a low oscillator frequency of only 168 K, as compared to the bulk BaTiO$_3$ Debye temperature of $\theta = 486$ K.

The quantitative analysis of the observed b_2^m crystal-field terms was rendered possible by the superposition model parameters for Cr^{3+} in octahedral oxygen coordination, as reviewed in Sect. 1.2.4. The b_2^m terms were well accounted for as far as sign and orientation of the magnetic axes are concerned. A satisfactory agreement regarding the magnitude could be reached by assuming the effective Cr^{3+}-O

Table 1.3. Comparison of Cr^{3+} EPR data in the three FEP's of $BaTiO_3$ with two models on the basis of the superposition model. From [1.70]

Phase	Experimental EPR data $(10^{-4}\,cm^{-1})$	One parameter model[1] $(10^{-4}\,cm^{-1})$	Centered model[2] $(10^{-4}\,cm^{-1})$
Tetragonal	$D_t^P = -199(2)$ $\mathrm{sgn}\,D_t^P = -\mathrm{sgn}\,D_o^P$	$D_t^P = -213(36)$	$D_t^P = -199$
Orthorhombic	$D_o^P = +123(1)$ $E_o^P = +32 - E_o^T$	$D_o^P = +108(18)$ $E_o^P = -98(17)$	$D_o^P = +115(19)$ $E_o^P = -101(17)$
Rhombohedral	$D_r^P = +60(1)$	$D_r^P = +83(14)$	$D_r^P = +229(39)$

[1] Deviation Δ the same as for Fe^{3+} in [1.14]
[2] D_t^P used to determine $|t_2|$.

distance to have shrunk by 0.02 Å from the intrinsic six oxygen positions towards the center. This can be regarded as a consequence of the large Cr^{3+} ionic fluctuation towards the oxygens, absent for Fe^{3+}. Thus, the absence of e_g electrons for Cr^{3+} in $BaTiO_3$ does indeed render the potential considerably flatter than that for Fe^{3+} in $BaTiO_3$, but the charge misfit suffices to keep the potential minimum at the center of the octahedron. Table 1.3 reproduces the measured and computed spin Hamiltonian constants. Note that the measured $E(T) = E_0^P + E_0^T = 32 \times 10^{-4}\,cm^{-1}$ requires an $E_0^T = 130 \times 10^{-4}\,cm^{-1}$ fluctuation term, in agreement with the computed $E_0^P = -101 \times 10^{-4}\,cm^{-1}$. This is reasonable because the deduced E_0^T is comparable to D_0^T in this phase. The former reflects thermal fluctuations along $\langle 001 \rangle$, the latter along $\langle 110 \rangle$ directions. These fluctuation amplitudes of Cr^{3+} in its octahedral oxygen cage therefore appear to be comparable along $\langle 001 \rangle$ and $\langle 110 \rangle$.

The t_2 exponent of the superposition model explaining the Hamiltonian parameters D^T in the tetragonal and orthorhombic phases of $BaTiO_3$ is $t_2 = +0.38 \pm 0.04$. It is of the same size as that obtained for $SrTiO_3$ but of opposite sign. For the latter perovskite, $t_2 = -0.36$ [1.14]. This proves that $R > R_m$ in $BaTiO_3$, where R_m is the R value for $\bar{b}(R)$ at the maximum. Thus the different sign of t_2 in $SrTiO_3$ is experimental proof that the maximum of $\bar{b}(R)$ is near 1.967 Å. The difference between R_m and R is about the same for $SrTiO_3$ and $BaTiO_3$ but in opposite directions, $|R - R_m| \simeq 0.015$ Å. On the other hand, were the oxygens surrounding the Cr^{3+} at their intrinsic distance $\bar{R} \simeq 2.003$ Å with $R_m = 1.967$ Å, then $(R - R_m) = 0.036$ Å, that is the data are consistent with the above-mentioned effective inward relaxation.

A number of related studies in other ionic ferroelectrics have been made, for instance, in boracites [1.71, 72], where metal site symmetries could be identified from Mn^{2+} spectra and a new structure for the monoclinic phase in Zn-Cl boracite was proposed. When the local point symmetry lacks inversion, large linear shifts of EPR lines have been observed as a function of external electric fields. An example is the Gd^{2+} on a D_{2d} site in tetragonal paraelectric $SrTiO_3$ [1.73, 74].

1.4.4 Mn^{4+} on Ferroelectric Ti^{4+} Sites: Evidence for Intrinsic Low-Frequency Dynamics

In the preceding section, it was shown that the Cr^{3+} EPR measurements give evidence that this ion remains centered in the oxygen octahedra (as does Fe^{3+}) when it is substituted for Ti^{4+} in $BaTiO_3$. As already mentioned, this is due to the repulsion by the negatively charged oxygens of the effective negative charge of the triply charged ions [1.39]. This is not the case for Mn^{4+} ions, for which theoretical calculations [1.75] indicate a ground-state energy similar to that of Ti^{4+}. Therefore, it could be anticipated that the Mn^{4+} might be off-center just as the Ti^{4+}.

To substantiate the theoretical predictions, EPR of Mn^{4+} in the rhombohedral low-temperature phase of $BaTiO_4$ was carried out [1.76], see Fig. 1.9. The axial crystal-field splitting $|D| = 0.65 \, cm^{-1}$ found for Mn^{4+} is much larger than that of Cr^{3+} in the rhombohedral phase, namely $D = 0.023 \, cm^{-1}$. One should note that when the two ions occupy the same position, as they do in corundum, the splittings D are almost the same [1.4]. By using the superposition model in its truncated version, the value of D can be estimated for different hypothetical positions of the substitutional ion. By comparing the values of b_2^0 for Cr^{3+}, Fe^{3+} and Mn^{4+}, agreement with the observed experimental data is obtained by letting the Mn^{4+} sit in an off-center position by $\simeq 0.14 \, \text{Å}$, near the intrinsic Ti^{4+} off-center position. Here we mention that *Bersuker*, from vibronic theory, correctly predicted that the Ti^{4+} ions lie off-center along $\langle 111 \rangle$ directions [1.77].

A remarkable feature of the EPR study is the absence of the Mn^{4+} EPR spectrum in the high-temperature phases of $BaTiO_3$ when measuring at 19.3 and 13.0 GHz. For this to occur, the Mn^{4+} has to reorient with time constants of the order of 10^{-9} to 10^{-10} s. If it reoriented faster, an average Mn^{4+} spectrum should be observed. No such spectrum has been detected when heating up to 800°C. Were the Mn^{4+} reorientation activated thermally according to an Arrhenius law, an averaged spectrum should have been seen. Thus the reorientation time in the high-temperature phase has to have a kT dependence that is much weaker than exponential. This argumentation concerns the EPR dynamics due to the secular terms in the Hamiltonian and the splittings for H directed near $\langle 111 \rangle$ where they are largest. For $\bar{H} \| \langle 100 \rangle$, the EPR lines for all different z-directions (parallel $\langle 111 \rangle$) coincide when the Hamiltonian is purely axial. However, there is a strong linewidth anisotropy $H(\theta)$ with lines much larger for H perpendicular to a certain z-direction than for H parallel to z. These anisotropies are partially due to secular terms of random amount proportional to $(S_x^2 - S_y^2)$, but nonsecular contributions also appear to be present. The latter are proportional to the spectral density of fluctuations $J(\nu)$ at frequency $\nu = g\beta H/\hbar$. This means that fluctuations with ν near 10^{10} Hz are important for the resonance fields H. The Mn^{4+} with nearly the same atomic mass as Ti^{4+} follows its cooperative motion and allows its dynamics to be probed. A local mode of the Mn^{4+} can be excluded with certainty because in the nonferroelectric $SrTiO_3$, the Mn^{4+} spectrum is visible between 4 and 300 K [1.78].

Since the EPR of Mn^{4+} is a reliable probe of the Ti^{4+} position and motion, this resolves an important aspect of the statics and dynamics of the ferroelectric phase transition in $BaTiO_3$. In their X-ray work, *Comes* et al. [1.79] had concluded that

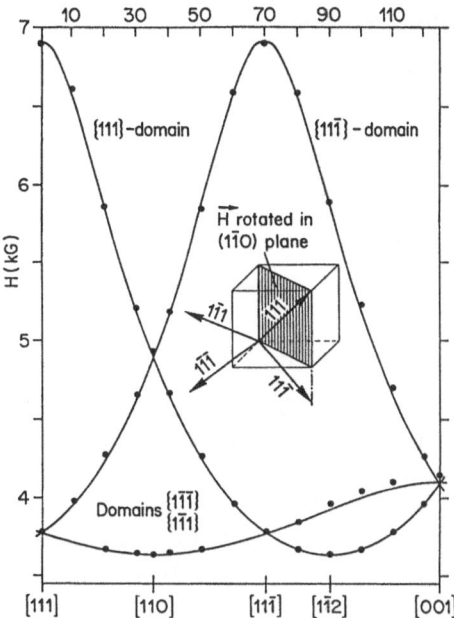

Fig. 1.9. Center of paramagnetic resonance fields of Mn^{4+} in rhombohedral $BaTiO_3$ at 4.2 K and 19.2 GHz. Lines from various domains are indicated. Note that EPR does *not* distinguish between $\{111\}$ and reversely polarized $\{\bar{1}\,\bar{1}\,\bar{1}\}$ domains, etc. Full lines are calculated theoretically. From [1.93]

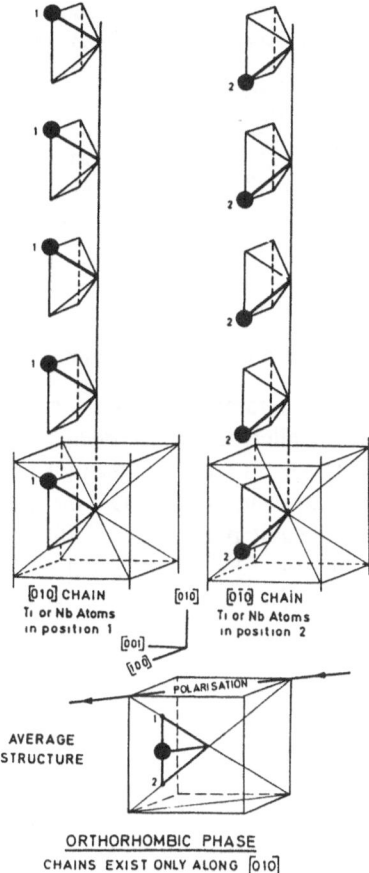

Fig. 1.10. The Ti or Nb displacements in the [010] and [0$\bar{1}$0] chains of the average orthorhombic structure. Reprinted with permission from [1.79]. Copyright 1968 Pergamon Journals Ltd.

the Ti ion sits off-center along equivalent $\langle 111 \rangle$ directions in *all* phases: at rest in the low-temperature phase, and motionally averaging over two equivalent $\langle 111 \rangle$ directions in the orthorhombic, four in the tetragonal, and all eight $\langle 111 \rangle$ in the cubic phase, see Fig. 1.10. The particular [111] displaced Ti ions are correlated along $\langle 100 \rangle$ equivalent directions. Owing to these rather long correlated chains, the reorientation has to be slow as deduced from the new Mn^{4+} EPR investigations [1.76]. Because this motion is so slow, of the order of 10^{-9} s, it escaped observation by neutron, Raman or equivalent scattering experiments but appears static to X-rays [1.79, 80]. The reorientation times cannot be much slower than the time scale set by the cubic crystalline splitting of Fe^{3+}, $\frac{1}{5}ac \simeq 10^{-9}$ s, because the EPR of Fe^{3+} is observed in *all* phases of $BaTiO_3$. This is shown schematically in Table 1.4.

Table 1.4. Schematic representation of observed and unobserved EPR of Mn^{4+} and Fe^{3+} in the four phases of $BaTiO_3$. From [1.93]

Ion	Characteristic time				
Mn^{4+}	10^{-10} s	observed	not observed	not observed	not observed
Fe^{3+}	10^{-9} s	observed	observed	observed	observed
		—————	————+————	————+————	————+————→ T
Phase		rhombohedral	orthorhombic	tetragonal	cubic

1.4.5 The EPR Parameter a of Fe^{3+} in $BaTiO_3$ and $KNbO_3$

The a parameters of Fe^{3+} shown in Table 1.1 for $BaTiO_3$ are about a factor of 2.5 smaller than those measured in normal cubic oxides with the same cation-anion distance, Fig. 1.11. This fact and its possible implication remained unnoticed for several years. In $BaTiO_3$, the soft ferroelectric mode is overdamped and anisotropic in the cubic phase, as in $KNbO_3$, which shows the same set of three phase transitions. Thus, the EPR of Fe^{3+} in $KNbO_3$ was investigated experimentally by *Siegel* et al. [1.68]. The only phase for which results are available so far is at room temperature,

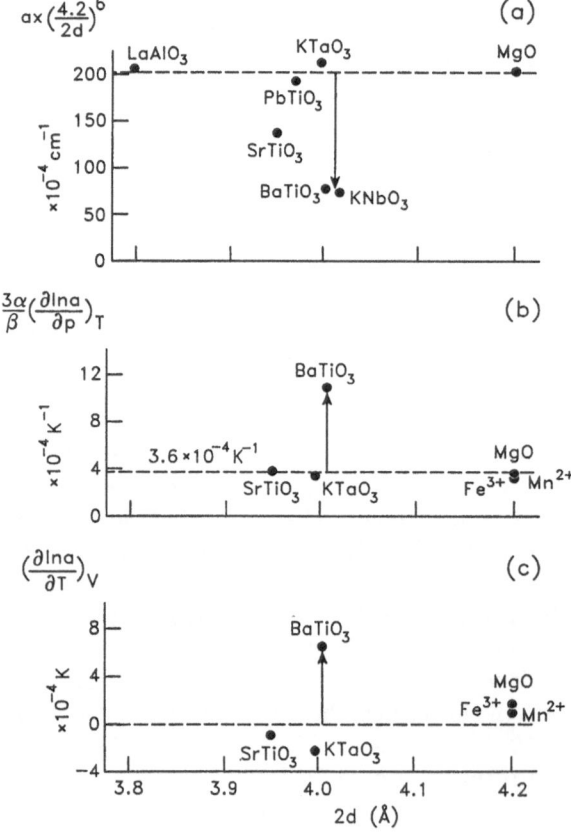

Fig. 1.11a–c. Comparison of scaled cubic field-splitting parameter a, relative explicit volume (reve) and temperature (rete) effects in various cubic oxides as a function of lattice parameter $2d$. From [1.85]

i.e., the orthorhombic phase. It turned out that a is essentially the same as in $BaTiO_3$. The measured parameters a in near-cubic oxides were shown to vary in a systematic way as a function of the intrinsic lattice constant d [1.81]. They follow an $a = a_0/d^6$ law irrespective of the nominal charge of the ion which the Fe^{3+} replaces. Figure 1.11a shows a values scaled by this law. The anomalously low value of a in $BaTiO_3$ and $KNbO_3$ is striking. It is further emphasized by EPR measurements of *Rytz* et al. [1.82] in mixed crystals, $KTa_{1-x}Nb_xO_3$ for $0 \le x \le 0.15$. Although the lattice constant d remains essentially the same, the EPR parameter a decreases linearly with x from the pure $KTaO_3$ value of $0.0305\,\mathrm{cm}^{-1}$ and extrapolates to $0.010\,\mathrm{cm}^{-1}$ for pure $KNbO_3$. Interestingly, the ferroelectric transition temperature increases linearly with x [1.83].

We can, with restrictions, take advantage of the empirical $a(d)$ dependence to obtain information about the local potential $V(R)$ in which the octahedral ion sits: We first compute, from $a(d)$, how much larger d_{eff} in $BaTiO_3$ has to be than its actual lattice constant d to observe a reduced by a factor of 2.5. From the power law dependence of a on d, we get $d_{\mathrm{eff}}/d = 2.5^{1/6} = 1.17$, i.e., the probing Fe^{3+} sees the oxygens in $BaTiO_3$ at a *distance 17% larger* than it would be for an inert oxide. Consider the anharmonicity of $V(R)$ with R parallel to [111] with $|R| \equiv R$. It can be parametrized as

$$V(R) = -AR^2 + BR^4 \quad , \tag{1.16}$$

with positive constants A and B. This potential has minima at $R_m = \pm\sqrt{A/2B}$ with energy $V_m = (A/2)R_m^2$. We now assume that the distance of the minimum R_m is also larger by the same factor $R_m(BaTiO_3)/R(IO) = d_{\mathrm{eff}}/d$, where I O stands for inert cubic oxide. Of course, $R_m \ne d$, but to lowest order their variation is proportional. With this, we calculate for $V_m = (A/2)R_m^2$ in $BaTiO_3$, $V_m(BaTiO_3) = 1.34\,V_m(IO)$, a 34% enhanced anharmonicity $\propto 1/B$. The same enhanced anharmonicity must also be present in $KNbO_3$. The distinction between the limiting cases of displacive versus order-disorder behavior at the transition temperature T_c is determined by whether $V_m \ll kT_c$ (displacive), or $V_m \gg kT_c$ (order-disorder) [1.84].

At this point, the question arises of whether additional experiments can confirm the conclusion reached above. This has indeed been true for recent measurements of the pressure and temperature dependences of the cubic crystalline splitting parameter $a(p, T)$ [1.85] whose total differential is given by

$$da = \left(\frac{\partial a}{\partial p}\right)_T dp + \left(\frac{\partial a}{\partial T}\right)_p dT \quad . \tag{1.17}$$

$(\partial a/\partial p)_T$ and $(\partial a/\partial T)_p$ were first measured for Fe^{3+} and Mn^{2+}; then using the differentiated form of the equation of state $V = V(p, T)$, where V is the volume, *Walsh* et al. [1.86] obtained the relation

$$\left(\frac{\partial a}{\partial T}\right)_p = -\left(\frac{3\alpha}{\beta}\right)\left(\frac{\partial a}{\partial p}\right)_T + \left(\frac{\partial a}{\partial T}\right)_V \quad , \tag{1.18}$$

where $\alpha = (1/d)(\partial d/\partial T)_p$ is the coefficient of linear thermal expansion, and

$\beta = -(3/d)(\partial d/\partial p)_T$ the volume compressibility. The first term on the right-hand side is the explicit volume effect, and the second the explicit temperature effect. $(\partial a/\partial p)_T$ and $(\partial a/\partial T)_p$ have also been measured for Fe^{3+} in $SrTiO_3$ [1.87], and more recently for $KTaO_3$ [1.88]. In the latter publication, *Rytz* et al. compared the values obtained for Fe^{3+} and Mn^{2+} in MgO as well as Fe^{3+} in $SrTiO_3$ and $KTaO_3$. Two very interesting properties of $a(p,T)$ were noticed. The explicit volume effect relative to a, reve, $(3\alpha/\beta)(\partial \ln a/\partial p)_T$, was, to within 6%, the same for all four measurements, see Fig. 1.11b. The optical modes for either MgO, $SrTiO_3$, or $KTaO_3$, being underdamped, indicate quite harmonic potentials. The explicit temperature effect is negative in $KTaO_3$ and $SrTiO_3$, whereas it is positive for Mn^{2+} and Fe^{3+} in MgO. Figure 1.11c shows the relative explicit temperature effect, rete, $(\partial \ln a/\partial T)_V$. It was pointed out in [1.88] that there are two contributions to the explicit temperature effect

$$\left(\frac{\partial a}{\partial T}\right)_V \simeq +f_1 - f_2 \frac{T_c}{(T-T_c)^2} \quad , \tag{1.19}$$

a positive Debye contribution $+f_1$ and a negative soft-mode contribution given by $-f_2 T_c/(T-T_c)^2$. The retes, i.e. $(1/a)(\partial a/\partial T)_V$ shown in Fig. 1.11c reflect this in a clearer way. The values for Mn^{2+} and Fe^{3+} in MgO are almost identical. In this crystal, the soft-mode term is absent, whereas it is present in $KTaO_3$ and $SrTiO_3$. The two rete values of $KTaO_3$ and $SrTiO_3$ are negative and comparable to each other.

In Figs. 1.11b and c, the recent reve and rete values for $BaTiO_3$ are included and marked by arrows from the average and zero lines, respectively. They were obtained from the pressure and temperature dependences of a in the cubic and tetragonal phases [1.85]. They show the following remarkable results: The *explicit relative volume effect is larger by a factor of 3.0* than for the other oxides, and confirms the 34% enhanced anharmonicity of the Ti ion deduced from its absolute 2.5 times reduced a value as compared to other oxides. The *relative explicit temperature effect* $(\partial \ln a/\partial T)_V$ *is positive, 4.5 times* that in MgO, whereas it is negative in $SrTiO_3$ and $KTaO_3$ with underdamped soft modes. This giant positive rete masks a possible negative soft-mode contribution of -2.0, and demonstrates the substantial order-disorder character of $BaTiO_3$ with its more anharmonic potential.

Anharmonic dynamical lattice theory explains the soft mode and dielectric constants in the incipient ferroelectrics $SrTiO_3$ and $KTaO_3$, for example, in the work of *Cowley* and *Bruce* [1.84], or more recently *Migoni* et al. [1.89]. However, the Fe^{3+}, Cr^{3+} and Mn^{4+} EPR findings make it questionable whether such a theory can quantitatively account for the three classic ferroelectric phase transitions in $BaTiO_3$ and $KNbO_3$. In fact, some dynamical results indicate that in $BaTiO_3$ and more clearly in $KNbO_3$ the soft mode rather freezes out towards the orthorhombic-*trigonal* transitions [1.90], i.e., $\langle 111 \rangle$ displaced Ti(Nb) ions are involved. The Fe^{3+} EPR results and also those on Mn^{4+} make the theories based on electronic bands and densities [1.91, 92] appear at least as promising [1.93].

1.5 Antiferrodistortive Transitions

Structural phase transitions in which the unit cell is at least doubled below T_c are termed antiferrodistortive [1.94], the most prominent examples of this variety occurring in $SrTiO_3$ and $LaAlO_3$. Both have been used as model substances for SPT, from the discovery of their transitions to the most recent results on critical and multicritical phenomena. EPR has thereby played an important role. The EPR spectra of Fe^{3+} ascertained the existence of a cubic-to-tetragonal transition near 100 K and domains below T_c in $SrTiO_3$ three decades ago [1.15]. It was again EPR, both in $LaAlO_3$ [1.95] in 1964 and in $SrTiO_3$ [1.96] in 1967, that allowed determination of the correct low-temperature space group. Furthermore, EPR *directly* disclosed the basic rotational properties of oxygen octahedra in these compounds, first in $LaAlO_3$ [1.95], then in $SrTiO_3$ [1.96]. By virtue of this property, the two phases are closely related, and we review these results in Sects. 1.5.1 and 1.5.2 [1.97, 98]. The temperature dependence of the rotational order parameter $\varphi(T)$ was measured in both compounds, and led to the first observation of static critical properties in SPT's [1.99]. This work is included here because an early summary [1.2] is long since out of print.

Subsequently, the importance of coupling the rotational order parameter to strains in the critical regime was recognized [1.100]. An example is the fluctuation-induced first-order transition that was shown to be restored to continuous by application of a symmetry-breaking uniaxial stress [1.101]. In Sect. 1.5.3, we summarize some important critical phenomena as studied by EPR, and finally, in Sect. 1.5.4, we review the most recent findings for Gd^{3+} in $PrAlO_3$ which elucidated the lowest-temperature SPT and the order parameter in the quantum regime.

1.5.1 The Cubic-to-Tetragonal Transformation in SrTiO$_3$

Both $SrTiO_3$ and $LaAlO_3$ crystallize in the ABO_3 perovskite structure as does $BaTiO_3$, and are cubic at sufficiently high temperatures, see Fig. 1.5. $SrTiO_3$ was thought to remain in this phase down to quite low temperatures, but in 1956 *Gränicher* [1.102] observed a slight anisotropy in the dielectric constant below 100 K. However, no splitting of X-ray lines was found. An analysis of the spectrum of Fe^{3+} substituting for Ti^{4+} was then carried out. In the cubic phase, the Fe^{3+} spectrum with $S = \frac{5}{2}$ could be described by the spin Hamiltonian (1.6), the resonance pattern reflecting the cubic environment of the Fe^{3+} ions [1.15]. Below 100 K, the fine structure $\pm\frac{1}{2} \leftrightarrow \pm\frac{3}{2}$ and $\pm\frac{3}{2} \leftrightarrow \pm\frac{5}{2}$ transitions were observed to split for magnetic-field directions along [100], but not for [111] directions. This necessitated the introduction of the axial term $D\left[S_z^2 - \frac{1}{3}S(S+1)\right]$ with $z \| \langle 100 \rangle$. It was concluded that $SrTiO_3$ undergoes a second-order phase transition near 100 K, and consists of tetragonal domains below this temperature [1.15]. These findings were confirmed by an American group [1.103] as well as by the EPR of Mn^{4+} ($3d^3$) on Ti^{4+} sites [1.78] and Gd^{3+} on Sr^{2+} sites [1.57]. It was noted in the first Fe^{3+} study [1.15] that the transition may be typical for oxides crystallizing in the ABX_3 perovskite structure. The latter can be regarded as composed of positive A ions and negative, near-rigid

BX$_6$ octahedra with common anion corners. Thus the octahedra can rotate about the B ion, the rotation angle being proportional to the oxygen displacements. In SrTiO$_3$ this rotation occurs around [100] axes and was reported by *Unoki* and *Sakudo* [1.96] (Fig. 1.12). One notes the correspondence with *Buerger*'s model of an ideal distortive transition proposed in 1950 [1.104]. For a particular domain, owing to the fourth-order term, there are two Fe^{3+} spectra which are rotated relative to each other by 2φ, the difference in TiO$_6$ sublattice rotations. Because the short-range character gives rise to the fourth-order term, the x, y and z directions in the Hamiltonian (1.2) point along the corners of the rotated octahedra in Fig. 1.12.

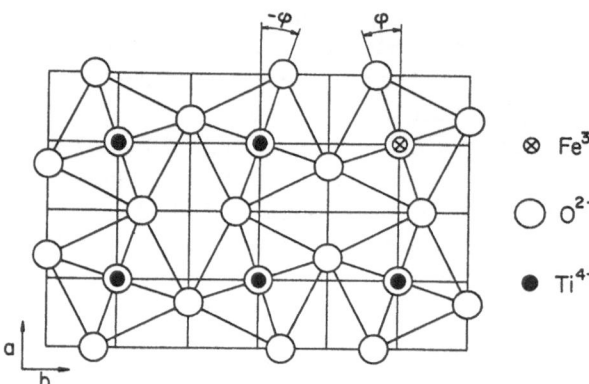

⊗ Fe^{3+}

◯ O^{2-}

● Ti^{4+}

Fig. 1.12. Oxygen octahedra rotated around the tetragonal c-axis in SrTiO$_3$ below T_c. The presence of an Fe^{3+} substitutional defect is indicated. From [1.1]

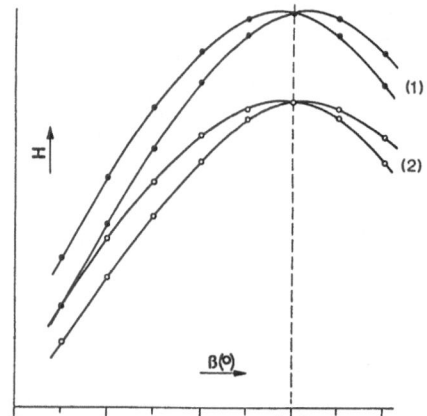

Fig. 1.13. Angular dependence of resonance magnetic fields of the $+1/2 \rightarrow +3/2$ (*1*) and $-3/2 \rightarrow -5/2$ (2) transitions in SrTiO$_3$ for *H* in a (100) plane at 77 K. From [1.2]

The angular dependences of the $+\frac{1}{2} \rightarrow +\frac{3}{2}$ and $-\frac{3}{2} \rightarrow -\frac{5}{2}$ transitions for a rotation of *H* in a (100) plane near a [100] direction are shown in Fig. 1.13 at a temperature of 77 K $< T_c$. From Fig. 1.13, it is clear that φ can be determined for various temperatures at constant magnetic field by rotating the magnet. The rotation from the line of one sublattice to the other then determines 2φ absolutely. This allows an accuracy of one tenth of an angular degree, the accuracy of the vernier of the magnet. By measuring the magnetic-field splitting of the two sublattice EPR lines at fixed angle near maximum slope of the resonance fields versus angle, a relative error of $\delta\varphi \simeq 1/100$ of an angular degree could be achieved. This corresponds to an accuracy in the position of the oxygen atoms of $\delta x = \delta\varphi \times d \cong 4 \times 10^{-4}$ Å, with d the titanium-oxygen distance $d = 1.95$ Å. This precision in δx was, at the time, more than one order of magnitude better than that of other methods.

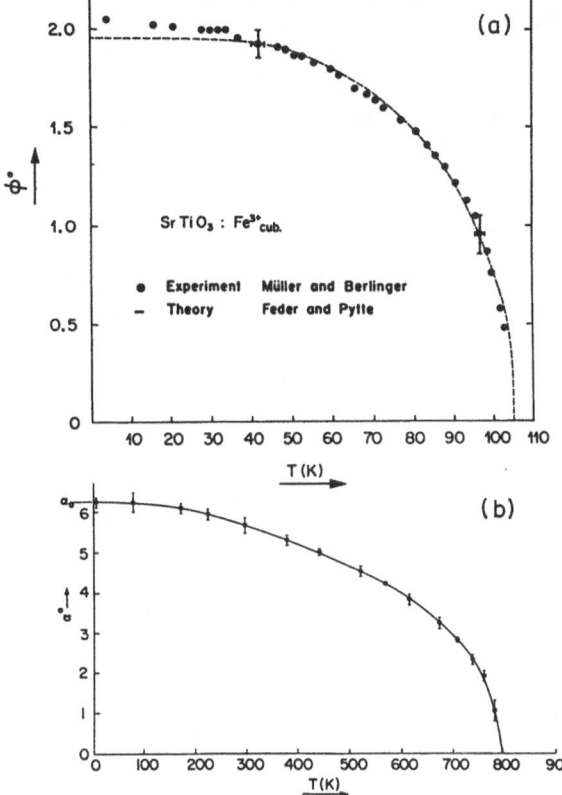

Fig. 1.14. (a) Temperature dependence of the rotational parameter φ in $SrTiO_3$. From [1.1]. (b) Trigonal rotation angle of the AlO_6 octahedra in $LaAlO_3$ as a function of temperature below T_a. From [1.97]

Figure 1.14a shows an earlier measurement of φ as a function of temperature and a fit to the microscopic theory of *Feder* and *Pytte* [1.105], which also accounts for the soft-mode results investigated by inelastic-neutron scattering by *Shirane* and *Yamada* [1.106] and Raman scattering by *Fleury* et al. [1.107] as well as elastic measurements [1.108]. It should be noted that, in the plane of rotation, the Ti-O distance for $T_c - 30 < T < T_c$ remains constant [1.109], and thus the rotation angle is the true order parameter of the transition.

From the Fe^{3+} spectra alone, it is not possible to determine the structure since the periodicity of the rotational motion in planes parallel to the one shown in Fig. 1.12 remained open. *Unoki* and *Sakudo* [1.96] pointed out that the EPR spectra of Gd^{3+} ions at Sr^{2+} sites [1.57] showed an axial second-order D term lying along the c-axis. This implies that octahedra on adjacent (100) planes rotate in opposite directions, i.e., if in one (100) plane an octahedron rotates by $+\varphi$ then the adjacent ones lying along [100] directions rotate by $-\varphi$. Thus from *EPR alone* it was possible to determine the low-temperature structure to be $D_{4h}^{18}, I/4mcm$. In the cubic phase, the periodicity of this staggered rotation is described by a wave vector $q = \frac{\pi}{a} \left(\frac{1}{2}, \frac{1}{2}, \frac{1}{2} \right)$ with a the lattice constant. q lies at the R corner of the Brillouin zone [1.106, 107]. The normal coordinate for this motion is therefore

$$\varphi_{[001]} = < \varphi > e^{-r_l q_R} \quad , \tag{1.20}$$

where r_l are Ti^{4+} lattice vectors. The rotation described so far around the [100] direction can occur for $T > T_c$ around any one of the [100], [010] and [001] directions. The three basis vectors $\varphi_{[001]}$, $\varphi_{[010]}$, and $\varphi_{[100]}$ for this motion transform as axial vector components. The mode is thus triply degenerate with R_{25} symmetry [1.105] and becomes dynamically unstable at T_c. ω_t is the soft mode of the transition.

1.5.2 The Cubic-to-Trigonal Phase Transition in LaAlO$_3$

a) **Experimental Results.** In this oxide, a rotation of alternate AlO$_6$ octahedra around pseudocubic body diagonals occurs as shown in Fig. 1.15. These $\langle 111 \rangle$ rotations can be described by a linear combination of the normal coordinates

$$\varphi_{[111]} = \frac{1}{\sqrt{3}} \left\{ \varphi_{[001]} + \varphi_{[010]} + \varphi_{[100]} \right\} \quad . \tag{1.21}$$

This shows how closely related the two low-temperature structures of SrTiO$_3$ and LaAlO$_3$ are. Four such combinations are possible, corresponding to rotations along the four pseudocubic body diagonals. Consequently in the low-temperature phase, four domains are observed, whereas in SrTiO$_3$ there are three (along equivalent $\langle 100 \rangle$ axes).

The alternate rotations were revealed by Fe^{3+} spectra at 300 K, and constituted historically the first observation of this characteristic phase transition possible in perovskite-type compounds [1.95]. Fe^{3+} ions are substitutional for Al^{3+} and require no charge compensation. To the spin Hamiltonian (1.6), with inequivalent x, y, and

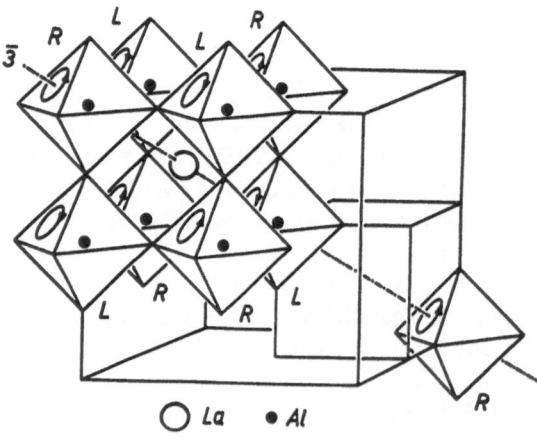

$\bigcirc$ *La* $\bullet$ *Al*

Fig. 1.15. Rotations of the AlO$_6$ octahedra under the $R\bar{3}c$ space group (for LaAlO$_3$). From [1.1]

z axes for two Al sites, uniaxial second- and fourth-order terms have to be added to analyze the data below T_c. The former is $D\left[S_\zeta^2 - \frac{1}{3}S(S+1)\right]$ for both Al^{3+} sites, and ζ is parallel to a pseudo-cubic $\langle 111 \rangle$ direction. The trigonal orientation of S_ζ^2, in contrast to the spin operators in the fourth-order terms which have components along quarternary axes, demands a different technique for the absolute measurement of the trigonal rotation angle. In $SrTiO_3$ this was achieved by a rotation of the external magnetic field in a (100) plane. Here, the trigonal axis of a monodomain sample is tilted by about 60° (for maximum sensitivity; the exact angle is not critical) to the magnetic field, and then the sample is rotated around itself. This procedure was first used by *Geschwind* [1.21] for trigonal Al_2O_3:Fe^{3+}.

EPR investigations of Gd^{3+} at La^{3+} sites [1.110, 111] and NMR quadrupolar splittings of ^{139}La [1.95], which indicate that the La lattice sites are equivalent, allowed an assignment to the $R\bar{3}c$ [1.22, 95] rather than to the $R3c$ structure based on alternate octahedral rotation alone. Earlier X-ray extinction could not discriminate between $R\bar{3}m$ and $R\bar{3}c$, but had shown this crystal to be rhombohedral from below about 700 K to 300 K [1.112]. In a hexagonal triple cell, for a rotation $\varphi_{[111]}$ of the AlO_6 octahedra, the oxygen parameter x is

$$x = \frac{1}{2}\left[1 \pm 3^{-1/2} \tan \varphi_{[111]}\right] \tag{1.22}$$

and agrees with that obtained from a neutron-diffraction study at room temperature [1.113]. An extended EPR investigation of Fe^{3+} determined $\varphi_{[111]}(T)$ as well as $D(T)$ from the phase transition at $T_c = 800 \pm 10$ K down to 4.2 K [1.97] where $\varphi_{[111]} = 6.2°$. This transition is the only one observed. T_c agrees with X-ray diffraction results [1.114] and was confirmed by differential thermal analysis (DTA) measurements [1.115].

The low-temperature structures of $SrTiO_3$ and $LaAlO_3$ retain their inversion symmetry. It readily follows that the EPR parameter D must in both cases depend quadratically upon the rotational parameter. In the classical Landau theory, the temperature dependence of the latter is proportional to $(T_c - T)^{1/2}$, thus $D \simeq (T_c - T)$. This linear dependence is observed over a large temperature interval in the experiments on $LaAlO_3$ except near T_c [1.95] (see also Fig. 1.17). The same behavior is also found for the NMR quadrupole splitting of ^{139}La and ^{23}Al [1.116] over that interval.

The static displacements in the $R\bar{3}c$ phase of $LaAlO_3$ are a linear combination of normal coordinates of $SrTiO_3$. Therefore the soft mode which freezes-out in the cubic phase is the same R_{25} mode. *Cochran* and *Zia* [1.117], in their analysis based on the $LaAlO_3$ EPR results, were the first to recognize the importance of this mode. It was detected by *Axe* et al. with inelastic neutron diffraction [1.118]. In the low-temperature rhombohedral phase where the modes are again split, *Scott* [1.119] observed them by Raman scattering. Away from the neighborhood of T_c, all experimental information is accounted for by the theory of *Pytte* and *Feder* [1.105].

b) Landau Theory. To close this section, we want to relate the $SrTiO_3$ and $LaAlO_3$ transitions and assess the microscopic theory mentioned [1.105]. We do so by using a free-energy expression based on crystal symmetry. Such a Landau expression can describe the nature of the transition and the soft modes [1.98, 120]. In a more recent form it is written as [1.100, 121]

$$E = E_0(T) + a(T) \sum_{i=1}^{3} \varphi_i^2 + b(T) \left(\sum_{i=1}^{3} \varphi_i^2 \right)^2 + c(T) \sum_{i=1}^{3} \varphi_i^4 \quad . \tag{1.23}$$

The indices $1, 2$ and 3 of the normal coordinates are short-hand notations for [100], [010] and [001]. In such a Landau expression, the coefficient $a(T)$ changes sign at T_c and varies linearly with T as $a(T) = \alpha(T - T_c)$. This is a direct result of the anharmonicity of the lattice. The coefficients $b(T)$ and $c(T)$ are not strongly temperature dependent. They vary by about 10% according to [1.105], and represent the isotropic and anisotropic parts of the anharmonic potential, respectively. $E_0(T)$ contains all other degrees of freedom of the lattice.

The equilibrium states of the crystal are obtained by minimizing $E(T)$ with respect to the φ_i, $\partial E / \partial \varphi_i = 0$. For $b(T) > 0$, a second- and for $b(T) < 0$ a first-order transition is found. For a constant $\varphi = (\varphi_1^2 + \varphi_2^2 + \varphi_3^2)^{1/2}$ and $c(T) < 0$, the fourth-order expression $c(T)(\varphi_1^4 + \varphi_2^4 + \varphi_3^4)$ has minima at points $\tilde{\varphi} = (\pm\varphi_{tt}, 0, 0)$, $(0, \pm\varphi_{tt}, 0)$ and $(0, 0, \pm\varphi_{tt})$ and maxima at $\varphi = (1/\sqrt{3})(\pm\varphi_{tr}, \pm\varphi_{tr}, \pm\varphi_{tr})$ with all sign combinations. Thus, depending on the sign of the fourth-order term $c(T)$, only tetragonal or trigonal (rhombohedral, $c(T) > 0$) symmetries, respectively, are compatible with this model. Thus SrTiO$_3$ and LaAlO$_3$ just realize the two possibilities. The minimization procedure yields $\varphi \propto (T - T_{tr})^{1/2}$ for the rotational parameter, with T_{tr} the transition temperature. The soft modes are found by solving the dynamical matrix

$$M\omega_i^2 \delta_{ij} = \frac{\partial^2 E}{\partial \varphi_i \varphi_j} \quad , \qquad (i, j = 1, 2, 3) \quad . \tag{1.24}$$

Coupling to other degrees of freedom, in particular static elastic deformations ε_{ij}, can be taken into account by adding coupling terms $\varepsilon_{ij}\varphi_i\varphi_j$ to (1.23) as well as the strain energy with terms of the form ε_{ij}^2 and $\varepsilon_{ij}\varepsilon_{lm}$. Minimizing E with respect to strains $\partial E / \partial \varepsilon_{ij}$ yields the strains as a function of the normal coordinates and the coupling constants involved. Thus, the strains can be eliminated and E renormalized to the form (1.23) with changed constants a', b' and c'. *Slonczewski* [1.122] considered the interaction with uniaxial stress in the phenomenological Landau model. To the renormalized Hamiltonian (1.23), two terms have to be added by symmetry. Neglecting hydrostatic pressure, these are

$$d_e \sum_i T_{ii}\left(3\varphi_i^2 - \varphi^2\right) + d_f \sum_{i<j} T_{ij}\varphi_i\varphi_j \quad , \tag{1.25}$$

where T_{ij} is the applied stress tensor.

The microscopic theory [1.105] yields the temperature-dependent constants of the Landau model. They are functions of the temperature-independent local octahedral second- and fourth-order potentials, the moment of inertia of the octahedra, the coupling between them and between the rotational and acoustic modes, and the rotational correlation functions which are temperature dependent. Comparison of both theories with static and dynamic experiments as a function of temperature places LaAlO$_3$ well in the trigonal phase, whereas SrTiO$_3$ is a borderline case between the tetragonal and trigonal phase. Therefore, trigonal-mechanical [111] stress can induce the $R\bar{3}c$ phase in $SrTiO_3$. We shall come back to this realization in Sect. 1.6 on multicritical points. Expressions (1.23) and (1.25) contain no spatial dependence $\varphi_i(x)$. In the microscopic theory it is in principle there, but for the correlation functions involved, averages over q space were taken. As we shall see, these φ of x dependencies are important for the critical phenomena and the fluctuations occurring near T_{tr}, as will be discussed in the next two sections.

1.5.3 Critical Phenomena

Displacive phase transitions including the ferroelectric ones were analyzed two decades ago with the help of Landau potentials [1.123], or with microscopic theories using the mean-field approximation [1.105]. Near the transition temperature T_c, both yield a square-root dependence (exponent $\beta = 1/2$) of the displacive order parameter as a function of temperature difference $|T - T_c|$ for $T < T_c$. The calculated susceptibility χ diverges as the inverse of the latter quantity (exponent $\gamma = 1$), and the specific heat shows a jump Δc_p (exponent $\alpha = 0$). Deviations from these laws have long been found experimentally for gas-fluid and magnetic systems as well as quantum fluids near the critical point terminating a first-order phase boundary, i.e., where a second-order transition occurs with $\beta \simeq \frac{1}{3}$, $\gamma > 1$, $\alpha \neq 0$ [1.124]. They result from correlated fluctuations of the order parameter, whereas the Landau potential is correct for uncorrelated fluctuations with a Gaussian distribution (Gaussian model) [1.125]. The deviation from Landau theory becomes appreciable when the length of the correlated fluctuations $\xi \propto \chi^{1/2} \propto (T - T_c)^{-1/2}$ exceeds the range λ of the forces. For ferroelectrics with their long-range forces, this was estimated by *Ginzburg* [1.126] to occur so close to T_c that no deviation is experimentally detectable.

As systems displaying critical behavior near SPT's, $SrTiO_3$ and $LaAlO_3$ are particularly good examples because

a) their antiferrodistortive rotational transitions, i.e., Brillouin-zone R-boundary soft modes, imply a cancellation of long-range axial dipolar electric forces. This would then be analogous to the known cancellation of dipolar magnetic forces in antiferromagnets [1.127, 128].

b) Both crystals undergo second-order phase transitions, whereas most displacive antiferroelectric transitions are first order.

c) With EPR, the accuracy attainable in measuring the rotational order parameter $\varphi(T)$ of $\frac{1}{100}$ of a degree on oxygen displacements of 4×10^{-4} Å was one order of magnitude more accurate than that of such techniques as X-ray or inelastic neutron scattering.

The outcome of the measurements is shown in Fig. 1.16a for $SrTiO_3$ where $\varphi(t)^{1/3}$ is plotted as a function of T for two data sets; one obtained for the "cubic" Fe^{3+}, the other for the $Fe^{3+}\text{-}V_O$ center (to be discussed later). An almost straight line was obtained for $\beta = 0.33$. From Fig. 1.16a, one sees that from $t = 1$ to $t = 0.94$ the first 18 points follow the straight line, but afterwards they deviate upwards (towards larger β).

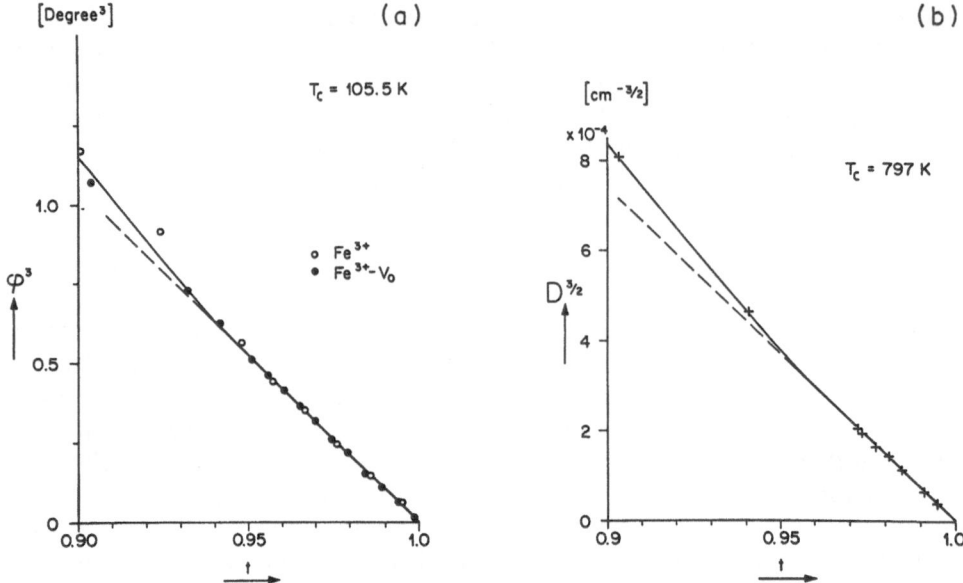

Fig. 1.16. (a) The cube of the rotational parameter φ versus reduced temperature $t = T/T_c$ in SrTiO$_3$. (b) EPR parameter $D^{3/2}$ versus reduced temperature t in LaAlO$_3$. From [1.99]

In LaAlO$_3$, determination of φ necessitates a rotation of the monodomain sample around the domain axis (inclined by $\simeq 60°$ to the magnetic field) and is far too inaccurate [1.97]. However, the crystalline field parameter D has been determined quite precisely. It reflects the local distortion of the octahedra along the trigonal axis. Near T_c, for small values, D must be proportional to φ^2 because the trigonal $R\bar{3}c$ structure has a center of inversion. Therefore, assuming $1/\beta$ close to 3, as in SrTiO$_3$, $D^{3/2}$ should be proportional to t. In Fig. 1.16b, this is seen to hold for the first seven points down to $t = 0.94$ with $T_c = 797$ K for our sample. A least-squares fit for these gave $1/\beta = 3.03$ [1.99]. X-ray intensity [1.114] and nuclear magnetic quadrupolar splittings of ^{23}Al and ^{139}La nuclei [1.116] are proportional to the EPR $D(T)$ parameter, and support the existence of a critical regime. They are, however, a factor of 3–5 less accurate.

To obtain the limit at which the behavior becomes classical, φ^2 of SrTiO$_3$ and D of LaAlO$_3$ were plotted as a function of t in Fig. 1.17. It is seen that a Landau-like straight-line behavior is approximately followed between $0.7 \leq t \leq 0.9$ in both systems. At $t_c \simeq 0.9$ or $1 - t_c \simeq 0.1$, the bending-down from the almost straight line becomes noticeable. There, the fluctuations $\varphi(t) - \langle\varphi(t)\rangle$ are of the order of the mean displacement $\langle\varphi(t)\rangle$ and have been estimated by Ginzburg [1.126] using the Landau approach and including a correlation term of the form

$$\sum_{i=1}^{3} (\nabla\varphi_i)^2 \quad .$$

From his work, one obtains the approximate relation

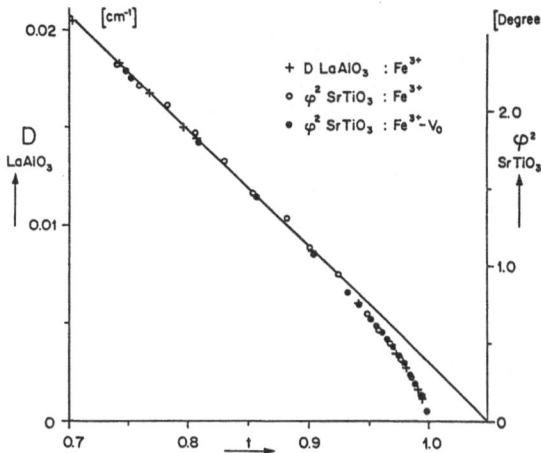

Fig. 1.17. φ^2 of $SrTiO_3$ and D of $LaAlO_3$ versus t between 0.7 and 1, showing the changeover from Landau to critical behavior. From [1.99]

$$1 - t_c \simeq \left(\frac{k}{\varrho \Delta c_p}\right)^2 \frac{1}{l^6} \quad , \tag{1.26}$$

where $\Delta c_p = 8.1 \times 10^{-4}$ cal/deg is the specific heat jump, if the system is described by the classical potential [1.120, 129] and $\varrho = 5.13\,\mathrm{g\,cm^{-3}}$. Equation (1.26) defines a zero-temperature coherence length l. One obtains $l \simeq 10$–$17\,\text{Å}$. As expected, this is of the order of the estimated range of forces, i.e. the distance between equivalent octahedral units of $8\,\text{Å}$.

It is remarkable (Fig. 1.17) how well the critical exponents coincide in magnitude for $SrTiO_3$ and $LaAlO_3$, despite one crystal being tetragonal and the other trigonal. The rotation angle at $T = 0$ is $2.1°$ in the former and $6.2°$ in the latter (Fig. 1.14). In addition, the transition temperature for $LaAlO_3$ ($T_c \simeq 800\,\text{K}$) is almost eight times higher than that of $SrTiO_3$ (105.5 K). Thus the behavior of the rotational parameters does indeed scale for these two second-order displacive transitions. Furthermore, φ^2 ($SrTiO_3$) and D ($LaAlO_3$) are proportional to each other to an appreciable extent, also outside the critical region.

These facts were not at first understood theoretically until it was recognized by *Aharony* and *Bruce* [1.100, 130] that they resulted from sample preparation. Proper shaping of crystals has led to strained samples which become almost monodomain below the phase transitions: to a $\{100\}$ domain in $SrTiO_3$ [1.131] and a $\{111\}$ domain in $LaAlO_3$. These samples allowed a much higher accuracy in the determination of $\varphi(T)$. Thus, the order-parameter dimensionality was $n = 1$ rather than $n = 3$ for a polydomain sample, and the *Landau-Ginzburg-Wilson* (LGW) Hamiltonian in renormalization-group (RG) notation from (1.23) for $i = 1$ including the gradient term reads [1.128, 130]

$$\mathcal{H}_1 = \frac{1}{2} \int \left[r_1(T)\varphi(\boldsymbol{x})^2 + \frac{u}{4}\varphi(\boldsymbol{x})^4 + c\left(\frac{d}{dx}\varphi(\boldsymbol{x})\right)^2 \right] dx \quad . \tag{1.27}$$

The renormalization of $\mathcal{H}_1$ yields the well-known value of $\beta_1 = 0.315$ and was within the error limits of the experiment, 0.32 ± 0.02. EPR linewidth anisotropy measurements confirmed the axial anisotropic Ising character of the fluctuation of the monodomain samples of $SrTiO_3$. Furthermore, the specific heat exponent for such samples was measured to be $0.25 < \alpha < 0.08$ and bracketed the Ising value of $\alpha = 0.125$ [1.132].

From Figs. 1.16 and 1.17, it is apparent that a crossing-over to mean-field behavior occurs in $SrTiO_3$ and $LaAlO_3$. This appears to be one of the very few instances where this has been observed for $d = 3$. A theory giving an analytical expression was only published later [1.133]. Renormalization-group theory yields a correction to scaling close to T_c of the form:

$$\varphi(T) = \varphi_0(1 - T/T_c)^{\beta_1} \left[1 + b_1(1 - T/T_c)^x \right] \quad . \tag{1.28}$$

Assuming $\beta_1 = 0.315$, a correction exponent $1 > x > 0.5$ was obtained [1.134], confirming recursion-relation calculations of $x \simeq 0.64$ [1.135] rather than others yielding $x = 0.5$ [1.136, 137].

Later, well-annealed and carefully etched polydomain $SrTiO_3$ samples were also measured with EPR; the second-order character of the transition was therewith confirmed [1.138]. These experiments gave a critical exponent of the order parameter of $\beta = 0.40 \pm 0.03$ [1.139], definitely higher than that found with monodomain samples and in agreement with the first RG results for SPT [1.140] obtained after the discovery of critical phenomena by EPR. These RG calculations used an $n = 3$ system appropriate for an unstrained crystal with a full cubic LGW Hamiltonian of the form:

$$\mathcal{H}_c = \int_x \left\{ \frac{1}{2} \left[r_0 \sum_{i=1}^{3} \varphi_i(x)^2 + \frac{u}{4} \left(\sum_{i=1}^{3} \varphi_i(x) \right)^2 \right] + v \sum_{i=1}^{3} [\varphi_i(x)]^4 \right.$$
$$\left. + c[\nabla \varphi(x)]^2 - f \sum_{i=1}^{3} \left(\frac{\partial \varphi_i}{\partial x_i} \right)^2 \right\} \quad . \tag{1.29}$$

The f term takes into account the cubic anisotropy of the fluctuations. These can be quite large in perovskites where octahedral rotations in one (100) plane (Fig. 1.12) are possible with hardly any correlation to the *next* (100) plane, in which case f approaches 1. It has been shown that for f not too close to 1, as is the case for $SrTiO_3$, and $v < 0$ from NMR results [1.141], f and u are irrelevant variables under renormalization, i.e., $f, v \rightarrow 0$, and a stable isotropic Heisenberg fixed point of the Hamiltonian with $u > 0$ exists for which $\beta \simeq 0.38$.

In Figs. 1.16a and 1.17, EPR rotation-angle measurements with the Fe^{3+}-V_O center are also shown. This center consists of a trivalent Fe^{3+} impurity substituting for a Ti^{4+} ion with a nearest-neighbor oxygen vacancy V_O [1.12]. The center is also known to exist in other ABO_3 crystals of the perovskite structure, and Me-V_O pairs have been shown to occur for other transition-metal Me^{2+} and Me^{3+} ions as well. Because of the oxygen vacancy, a strong axial crystal field exists at the Fe^{3+} site whose axis is directed along the Fe^{3+}-V_O pair-center axis. Thus a large EPR D term of $1.4\,cm^{-1}$ results. Using the Newman superposition model [see (1.8)], the local shifts of the ions could be inferred. It was shown that the Me ions (Fe^{3+} or Mn^{2+})

move by about 0.2 Å towards the oxygen vacancy [1.63]. The large D term yields a much stronger anisotropy of the EPR lines than the Fe^{3+} cubic center *for rotations φ perpendicular to the* Fe-V_O *axis* [1.63]. From Fig. 1.17, one sees that the intrinsic rotation φ and that of the Fe^{3+}-V_O center are closely proportional to each other. Thus the Fe^{3+}-V_O center does indeed represent intrinsic local rotational properties. The proportionality constant is given by $\varphi = (1.59 \pm 0.02)\bar{\varphi}$, i.e. $\bar{\varphi}$ is smaller than the intrinsic φ due to the local shift of the ions. This center was first employed to detect fluctuations in φ and to determine their anisotropy.

It was originally intended to include here a section on fluctuations and nonlinear phenomena as probed by EPR, with particular reference to the central peak phenomena. However, this has since been published, first in two lectures given in Norway [1.142] and then as an update in a more general context [1.143].

1.5.4 Order-Parameter Behavior in $PrAlO_3$

In the article of *Fleury* and *Lyons* in Chap. 2 of Vol. I of this series [1.144], the structural phase transitions in $PrAlO_3$ were presented. Since the writing of their article in 1980, EPR has been able to clarify the behavior of the rotation axis of the AlO_6 octahedra below 120 K. Therefore we end this section by summarizing these findings [1.145].

Below room temperature, $PrAlO_3$ undergoes several phase transitions not found in other perovskite compounds. In particular, the transitions at 205 and 151 K involve directional changes of the rotation axis of the AlO_6 octahedra towards different cubic axis, and result from the competition between R_{25}-phonon–Pr^{3+}-$4f$-electron interactions [1.146]. Figure 1.18 illustrates the sequence of structural phase transitions in $PrAlO_3$. As in $LaAlO_3$, a second-order transition from a cubic perovskite to a trigonal phase occurs at 1320 K due to the condensation of the R_{25} zone-boundary phonon mode. This phase persists down to 205 K, and the AlO_6 octahedra are rotated

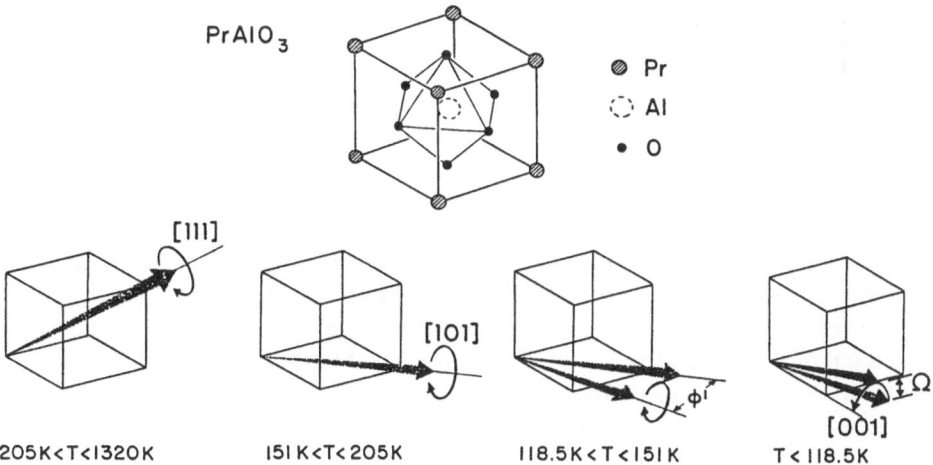

Fig. 1.18. The perovskite unit cell and the direction of the rotation axis of the octahedra in the various phases of $PrAlO_3$; $\varphi' = \pi/4 - \varphi$ and $\Omega \neq 0$ for $T < 118.5$ K. From [1.145]

by an angle $\simeq 9°$ about a [111] direction in a staggered fashion which is constant below 300 K. Cooling past 205 K, a first-order transition to an orthorhombic structure occurs, and the rotation axis for the AlO_6 octahedra jumps to a [101] direction. A second-order transition to a monoclinic structure takes place at 151 K, and the rotation axis moves continuously from the [101] towards the [001] direction below that temperature. This transition is a prototype of the cooperative Jahn-Teller effect involving a single electronic mode [1.147].

Some experimental evidence for another transition at 118.5 K has been reported in Brillouin scattering experiments [1.148], in Raman studies [1.146], and in EPR measurements [1.149]. On the basis of their data *Fleury* et al. [1.148] conjectured that the driving mechanism for this transition was the softening of a transverse acoustic phonon with strain as the sole order parameter. *Harley* subsequently proposed [1.146b] that the data could be explained by a linear coupling of an acoustic mode to the temperature-dependent optical modes in question. Below 151 K, the rotation axis of the AlO_6 octahedra lies in the (010) plane. In Harley's model, this axis would lift out of the (010) plane by an angle Ω below 118.5 K but would continue to approach [001] asymptotically. A confirmation of Harley's suggestion for the mechanism of the 118.5 K transition itself was presented using EPR techniques of Gd^{3+} ions substituted for Pr^{3+} in $PrAlO_3$, but the asymptotic behavior for $T \to 0$ is unexpectedly different [1.145].

The Hamiltonian of the Gd^{3+} ion used has been given in (1.6) and (1.14). Out of the seven transitions of the Gd^{3+} in $PrAlO_3$, the $-\frac{5}{2} \to -\frac{3}{2}$ resonance line was used to determine the order parameter, which is the *direction* of the axis about which the octahedra rotated. This direction is expressed in terms of the angle φ, varying between 45° and 0° as the rotation axis swings from the [101] towards the [001] crystal direction in the monoclinic phase in the (010) plane. The magnetic field was rotated in this plane and the EPR Hamiltonian z-axis then swings by an angle θ from 0° to 45° according to $\cot 2\theta = 0.80 \tan 2\varphi$ as found by *Sturge* [1.149]. The $-\frac{5}{2} \to -\frac{3}{2}$ resonance line was chosen for its sensitivity to angular variation of the field and the ability to follow and distinguish the line from the other resonances as the temperature and field position are varied.

Between 151 and 118.5 K, a characteristic splitting of lines called "φ" was observed due to (010) in-plane domains. An unsplit line is observed for domains whose octahedral rotation direction lies in (100) and (001) planes. Upon cooling below 118.5 K, this line was found to split by an amount "Δ". Because the "φ" lines remained unsplit, symmetry arguments indicate that the "Δ" splitting observed for (100) and (001) in-plane domains is due to a tilting of the octahedral axis out of these planes and not to a deformation of the entire crystal as proposed by *Fleury* et al. [1.148]. This conclusion was verified by setting the magnetic field parallel to the [001] direction, the crystal was then rotated in such a way that the magnetic field moved out of the (010) plane, in which it was rotated before, i.e., it was now moved in a (100) plane. In doing so, no splitting of the degenerate "φ" lines was observed, as a deformation of the entire crystal order would have caused. Only a displacement in the magnetic field was detected. This confirmed the Harley model.

Near $H \parallel$ [101], the "Δ" lines are almost parallel to each other as a function of H, and the *angle* Δ of the z-axis of the Hamiltonian measured from the [0$\bar{1}$1]

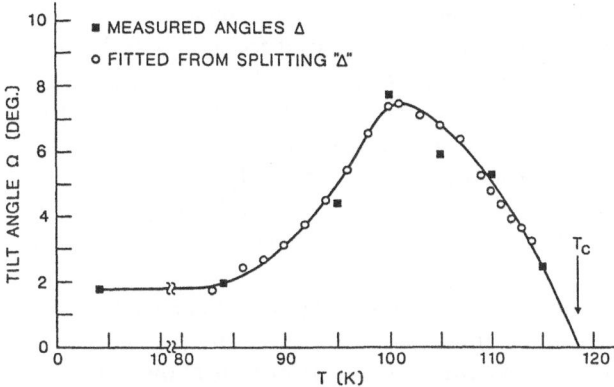

Fig. 1.19. Temperature dependence of the order parameter as obtained from the tilt-angle measurement (■) together with the fitted splitting (○) from the "Δ" lines below the lowest phase transition at 118.5 K in PrAlO$_3$. Note the finite value of 1.9° in the quantum region. The line is guide to the eye. From [1.145]

direction could be extracted. Then using the relation

$$\cot 2\Delta = 0.80 \tan 2\Omega \quad ,$$

analogous to what *Sturge* [1.149] found to hold between the Hamiltonian z-axis angle θ and the order-parameter angle φ *above* 118.5 K, Ω was computed from these measurements. Ω is shown by squares in Fig. 1.19. The circles have been scaled from the *magnetic* "Δ" splittings to which they are proportional. The rotation axis gradually tilts out of the (010) plane below 118.5 K down to $\simeq$ 100 K, and goes back towards the (010) plane with a further decrease in temperature without achieving a complete recovery at 4.2 K. Because the tilt angle $\Omega(T)$ becomes temperature independent below 80 K, one is most probably in the quantum regime of the system. Thus the PrAlO$_3$:Gd^{3+} study not only verified the Harley model for the 118.5 K transition of a tilt of the AlO$_6$ rotation axis out of the (010) plane, but also yielded unforeseen behavior in the quantum regime [1.145].

1.6 Experiments on Multicritical Points

1.6.1 Introduction

Multicritical points (MPC's) with different topologies can occur for a variety of reasons. Consider the isotropic Landau-Ginzburg-Wilson Hamiltonian $\mathcal{H}$ derived from (1.29) by putting $v = f = 0$. If u vanishes, a tricritical point results, and the nonvanishing higher-order terms, h.o., then stabilize $\mathcal{H}$. Another class of multicritical points will occur if the symmetry of $\mathcal{H}$ is broken, for example, by a uniform external field in an antiferromagnetic system, as emphasized at an early stage by *Fisher* et al. [1.150, 151].

At a tricritical point, a line of first-order transitions changes character to second order. A mixed-crystal $KTa_{1-x}Nb_xO_3$ study by *Todd* showed for $x = 0.30$ the first clear evidence of a para-ferroelectric tricritical point [1.152]. Upon application of hydrostatic pressure, this has also been shown to occur in the KH_2PO_4 ferroelectric crystal by *Schmidt* et al. [1.153]. The tricritical behavior can be accounted for by mean-field theory in three lattice dimensions, $d = 3$, especially in ferroelectrics because of the large range of dipolar forces and in part due to coupling between acoustic and optical modes. On the other hand, if the forces are short range, clear deviations from mean-field behavior will be observed near multicritical points of structural phase transitions. These result from breaking the symmetry by uniaxial stress as predicted by *Bruce* and *Aharony* from renormalization-group calculations [1.130]. EPR has concentrated on such experiments which in turn allowed one to check the RG results.

The predictions and experiments to be summarized here are concerned with the SPT's of ABX_3 compounds crystallizing in the perovskite structure. The first crystals investigated were $SrTiO_3$ and $LaAlO_3$, traditionally model substances for SPT's, see Sect. 1.5. Since the high-symmetry phase is cubic rather than isotropic in (1.29), v and f can be different from zero. A stress p can only couple *quadratically* to the order parameter because below the phase transition, the crystals retain their inversion symmetry. The symmetry-breaking term $\mathcal{H}_{sb}$ thus has the form $p_{ij}\varphi_i\varphi_j$, see especially Sect. 1.5.2b, and the total Hamiltonian is

$$\mathcal{H}_t = \mathcal{H}_c + \mathcal{H}_{sb} = \mathcal{H}_c + \sum_{ij=1}^{3} p_{ij}\varphi_i\varphi_j \quad . \tag{1.30}$$

For example, for a uniaxial stress p along [100], the last term in this expression becomes $g\left[\varphi_1^2 - \frac{1}{2}\left(\varphi_2^2 + \varphi_3\right)^2\right]$ with 1, 2, 3 meaning [100], [010], and [001], respectively, and $g = Cp$ [1.130].

In the following sections, multicritical points observed under high symmetry stresses in the two oxides mentioned as well as in $RbCaF_3$ and $KMnF_3$ are reviewed. We begin with the easiest case, the bicritical one. Here, two second-order critical lines meet at the MCP. Bicritical points have been observed in $SrTiO_3$ and $LaAlO_3$ by applying uniaxial stress along their respective [100] and [111] crystal axes. In Sect. 1.6.3, tetracritical behavior in $LaAlO_3$ and the Potts transition in $SrTiO_3$ are summarized. These result from applying tetragonal stress to $LaAlO_3$ and trigonal stress to $SrTiO_3$, i.e., with the stress symmetry axis reversed compared to the experiments described in Sect. 1.6.2. Section 1.6.4 reviews the most sophisticated MCP behavior realized: a tricritical Lifshitz point in $RbCaF_3$. Finally, the first observations of a critical end point in $RbCaF_3$ are discussed. The latter occurs when a second-order phase boundary ends on a first-order line.

1.6.2 Bicritical Points in SrTiO₃ and LaAlO₃

In $SrTiO_3$, the rotation of oxygen octahedra induces an elongation along the tetragonal c-axis below the phase transition [1.1, 109]. Thus, pulling along a particular [100] direction or pushing along a [011] axis favors this particular {100} domain over the other two as observed quite early [1.131]. Because of this coupling $T_c(p)$ is enhanced. Pushing along [100] favors the two {010} and {001} domains, again enhancing $T_c(p)$. In mean field, this shift is linear in p and the couplings c_{ij} in (1.30) were determined from ultrasound measurements [1.154]. The two phase boundaries are second order and merge in a bicritical point as emphasized by *Bruce* and *Aharony* [1.130], who also considered the critical regime close to T_c. They predicted a tangential merging of the boundaries $T_c(p)$ with a departure from linearity, according to

$$T_c(p) = T_c(0) + W p^{1/\phi} \quad , \tag{1.31}$$

where ϕ is the so-called shift exponent. For the lines separating the cubic from the two low-temperature phases, the exponent is $\phi = 1.25$ from scaling. Qualitative EPR Fe^{3+} linewidth measurements confirmed the bicritical regime [1.134]. Later, more quantitative ultrasound experiments by *Rehwald* [1.155] and accurate specific-heat measurements of $T_c(p)$ for (100) stress by *Stokka* and *Fossheim* [1.156] yielded $\phi = 1.27 \pm 0.06$, in very good agreement with the prediction. Figure 1.20a shows the Norwegians' results.

In $LaAlO_3$, rotation of the octahedra entails a rhombic angle larger than 90° [1.110], i.e., $LaAlO_3$ *shrinks* along the trigonal axis below T_c. Thus compression along a [111] direction yields one {111} domain and an $n = 1$ Ising line. Pulling leaves three equivalent {111} domains whose order parameters are linearly dependent upon one another. The resulting second-order boundary is an $n = 2$ XY line

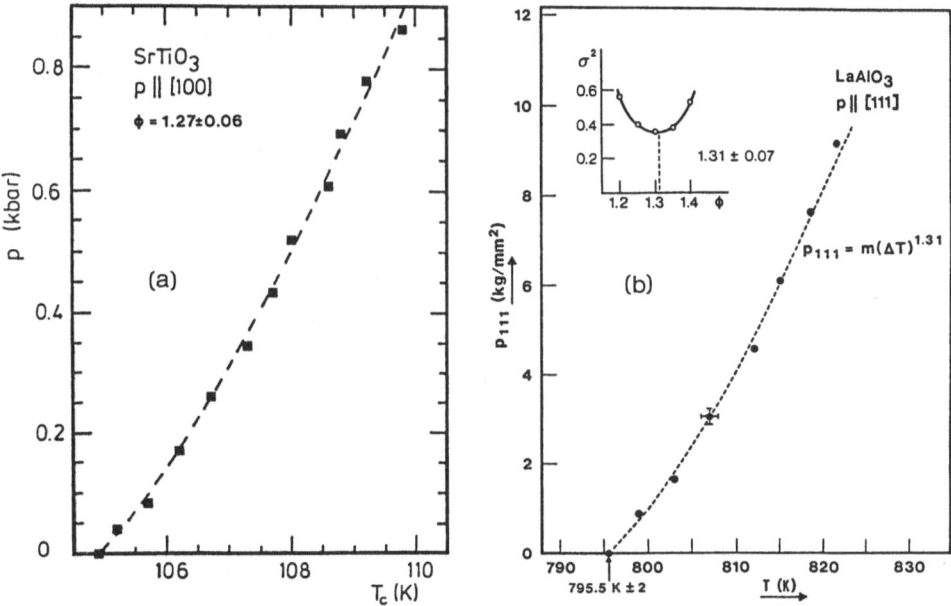

Fig. 1.20a,b. Phase diagrams of $SrTiO_3$ and $LaAlO_3$ with uniaxial pressure along [100] and [111], respectively, showing T_c versus p (dashed lines). ■ and ● are measured points and both dashed lines are the least-squares fitted curves to (1.31). The inset in (b) shows the standard deviation δ^2 as a function of exponent ϕ with a minimum at 1.31 ± 0.07. (a) from [1.156] and (b) from [1.157]

like that of SrTiO₃. Because T_c of LaAlO₃ is near 800 K, application of stress was not an easy task, especially as the accuracy in temperature had to be of the order of 10^{-5}. An EPR experiment using Cr^{3+} as a symmetry probe on Al^{3+} lattice sites determined $T_c(p_{111})$ for compressional p_{111} stresses [1.157]. The cavity especially developed for this purpose, see Fig. 1.4, is discussed in Sect. 1.3.2. The result is shown in Fig. 1.20b, again for the compressional branch of the diagram. The shift exponent determined in this manner was $\phi = 1.31 \pm 0.07$ in gratifying agreement with the SrTiO₃ results. Thus, one can assert that the bicritical behavior is well understood under conditions where the compressional stress is applied along the low-temperature domain axes.

1.6.3 Phases Induced by Competing Forces

Upon application of uniaxial stresses with a symmetry different from that of the genuine low-temperature phase of a system, other and new phases are induced. An experiment of this kind was actually the very first of all symmetry-breaking ones carried out; it involved [111] stress applied to SrTiO₃. In the Hamiltonian (1.29), v for SrTiO₃ is negative but small [1.122]. Thus a trigonal LaAlO₃-type phase was expected to occur from mean-field theory (MFT) and was indeed observed. In Fig. 1.21, the phase diagram as probed by the EPR of Fe^{3+} [1.158] is reproduced. Without uniaxial p_{111} stress, the known cubic O_h^1 to tetragonal $I4\,\mathrm{mcm}$ transition takes place upon cooling. For modest $p_{111} \neq 0$, a second-order transition to the trigonal LaAlO₃ phase is observed with monodomain, $n = 1$, character. Upon further cooling, the crystal wants to become tetragonal and a first-order transition to domains with components $\{xyy\}$, $\{yxy\}$ and $\{yyx\}$ of the rotation vector is observed. This phase diagram was accounted for by an independently determined Landau potential outside the critical regime [1.158]. Then it was recognized by *Aharony* et al. [1.159] that the first-order line was a realization of a three-state Potts transition not previously known in nature. This is because the three above-mentioned domains are reorientable by 120° rotations around the stress axis, but cannot be *reversed*.

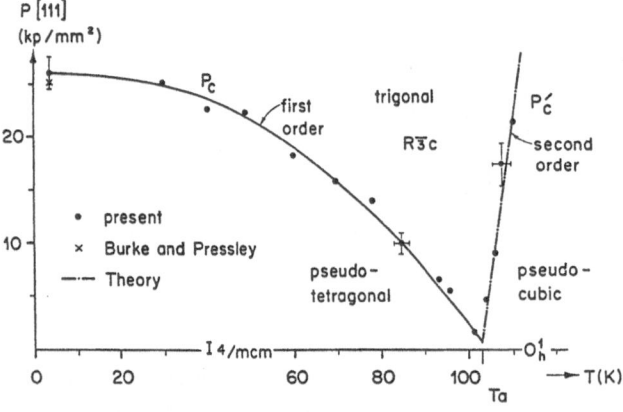

Fig. 1.21. Phase diagram of SrTiO₃ for the stress-induced $R\bar{3}c$ phase. For $T < T_a$ the transition is first order, for $T > T_a$ second order. From [1.158]

RG theory predicted the jump in order parameter $\delta\varphi$ at the boundary to be strongly renormalized by critical fluctuations with $\delta\varphi \propto (\varphi_{111})^{\delta^*}$ and $\delta^* = 0.56$ [1.160], as compared with the MFT $\delta^* = 1$ value. The experiment yielded $\delta^* = 0.62 \pm 0.06$, and can indeed be considered satisfactory.

Whereas the trigonal $R\bar{3}c$ phase in $SrTiO_3$ results from compressing the crystal along [111], the tetragonal $I4mcm$ phase in $LaAlO_3$ should occur under tensile stress along a [100] direction [1.130]. However, this has to be done at 800 K and has not yet been carried out. On the other hand, a {110} phase was predicted for compression parallel to [001]. In this phase, the AlO_6 octahedra rotate around the [110] direction. EPR of the Fe^{3+} substitutional on Al^{3+} sites verified this as correct. The phase diagram from this study is reproduced in Fig. 1.22 [1.157]. A second-order line T_1 separates the cubic ($\varphi = 0$) from the {110} phase. (It was the first second-order transition to a {110} phase ever reported.) This phase is adjacent to an intermediate phase with $\{yyx\}$, $x < y$, order-parameter components. At the T_2 phase boundary shown, the y-component continuously increases from zero. Below T_c and for zero stress, $x = y$, i.e., one is in the trigonal $R\bar{3}c$ phase described earlier. For tensile stresses $x > y$, the intermediate phasse continues until at a transition T_2', $y = 0$ and the {001} phase is reached. The latter vanishes at a second-order boundary $T_1' > T_c$ not shown. Thus, four phases and four second-order lines meet at the MCP, i.e., one has a tetracritical point of which the upper part with $p_{001} > 0$ has been experimentally verified. The precision of the data which served to construct the diagram of Fig. 1.22 was much less accurate than that of those shown in Fig. 1.20b, but later unpublished experiments have clearly revealed the critical behavior of T_1 and T_2 near the MCP.

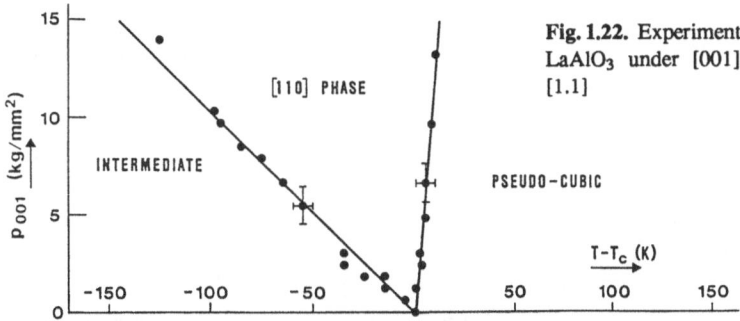

Fig. 1.22. Experimental phase boundaries of $LaAlO_3$ under [001] stress. Adapted from [1.1]

Data on multicritical behavior of $[1\bar{1}0]$-stressed $LaAlO_3$ have also been obtained [1.161] and so far indicate bicritical behavior, whereas according to *Blankschtein* and *Aharony* [1.162], an (001) phase separates the cubic from the intermediate phase, see the inset of Fig. 1.23. The experimental phase diagram consists of a second-order boundary separating the pseudo-cubic ($\varphi = 0$) phase from a phase with two $\{\bar{y}yx\}$ and $\{\bar{y}y\bar{x}\}$-type domains lying in the $(1\bar{1}0)$-stress plane. The multicritical point thus appears as bicritical rather than as tetracritical in Fig. 1.23, possibly because the intermediate (001) phase, i.e. $y = 0$, is rather narrow and so far has escaped detection. Also shown is the mean-field extrapolated temperature T_0 exceeding T_c

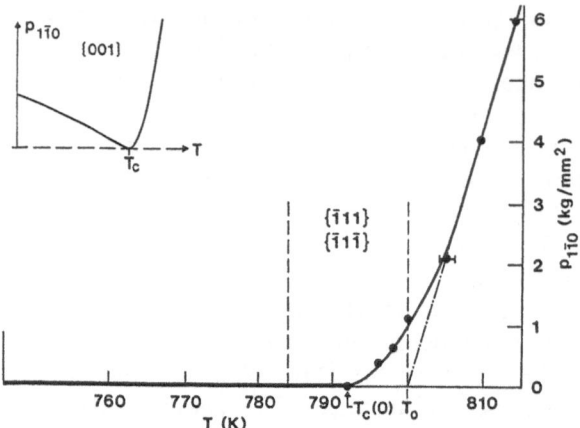

Fig. 1.23. Bicritical point of $LaAlO_3$ under $[1\bar{1}0]$ stress: first-order boundary for $p_{1\bar{1}0} = 0$ and second-order boundary for $T_c(p_{1\bar{1}0})$. Inset shows the predicted tetracritical point for Heisenberg fixed-point behavior. From [1.161]

by about 8 K, so that the critical region extends ± 8 K around T_c as indicated by the dotted lines. The critical region is $[T_c(0) - T_0]/T_c(0) \simeq 10^{-2}$.

1.6.4 Uniaxial Stress Effects on the Antiferrodistortive Transition of $RbCaF_3$ and Multicritical Behavior

a) Local Order Parameter Measurements in AMF_3 Compounds. For the SPT in $RbCdF_3$ ($T_c = 123$ K), an efficient EPR probe derived from the Fe^{3+}-V_O pair in $SrTiO_3$ was found to be the Fe^{3+}-O^{2-} pair. Its excellent sensitivity [1.163] is hampered by a SHF (^{19}F) structure resulting in an appreciable complication of the EPR lines. On the other hand, the ^{19}F interaction allows the local order parameters to be measured up to the third shell of fluorine ions surrounding $(3d)^5$ probes by means of ENDOR spectroscopy. This was done in $RbCdF_3$ doped with Mn^{2+} and Fe^{3+} [1.164]. In this way, using local measurements and static atomic displacements below T_c, it was shown that (001) layers of fluorine octahedra are weakly correlated in [001] tetragonal domains. Indeed, the local distortion of $\varphi(001)$ by the probe is not transmitted much to the nearest octahedron along [001]. Instead, the Gd^{3+} ion ($4f^7$), $S = \frac{7}{2}$, exhibits a weak SHF interaction giving rise only to a limited inhomogeneous line broadening. Moreover, its size nicely fits the host at a Ca site. In $RbCaF_3$ doped with Gd_2O_3, both cubic Gd^{3+} centers and Gd^{3+}-O^{2-} pairs are present, allowing two strategies to monitor the order-parameter behavior [1.13, 165]. It should be noted that the Gd^{3+}-O^{2-} pair is charge neutral in the $RbCaF_3$ lattice.

With the cubic center, the shifts of the fine-structure lines are essentially due to the axial b_2^0 term: $\Delta H \propto b_2^0(T) \propto \varphi^2(T)$ for $H \| [100]$ and for (100) tetragonal domains. The cubic centers are well suited for uniaxial stress (001) measurements with $H \| [100]$ or [010].

With the Gd^{3+}-O^{2-} pairs, the lines involving $M_s = \pm\frac{1}{2}$ transitions can be analyzed at X-band frequencies in terms of an effective g tensor: $g_\perp \simeq 8$, $g_\| \simeq 2\,(S = \frac{7}{2})$

instead of $g_\perp = 6$ and $g_\parallel = 2\,(S = \frac{5}{2})$ for the Fe^{3+}-V_O pair in $SrTiO_3$. The Gd^{3+}-O^{2-} center permits very accurate measurements of φ_{loc} for $H\|([110]\pm 2\varphi_{loc})$ and for $\langle 001\rangle$ domains, and it is well suited for measurements under uniaxial [110] and [111] stresses, with conveniently shaped samples (110 and 111 cylinders).

Details about the spin-Hamiltonian parameters of Gd^{3+} centers in AMF_3 crystals have been discussed [1.30, 166] in the light of the superposition and of the electrostatic models. As for Fe^{3+} in $SrTiO_3$, the local order parameter $\varphi(001)$ was also measured through the fourth-order terms $b_4^4 O_4^4$ using either the overcompensated Gd^{3+}-V_{Cd}^{2+} center in $RbCdF_3$ [1.167] or the cubic centers. In $TlCdF_3$, the Gd^{3+}-O^{2-} pair [1.168] permits EPR measurement at X-band frequencies near a spin level crossing, allowed in the high-temperature phase. The symmetry breaking at T_c then results in a sharp rise of a new line through level decrossing. The intensity of this line enables one to monitor the critical behavior of φ^2, just as the superstructure Bragg lines do for scattering experiments. Let us also note that for the SPT of $CsCaCl_3$ the table of ionic radii suggests the use of the Gd^{3+}-S^{2-} probe. This was recently carried out [1.169]. Whatever the particular methodological aspects of the Gd^{3+} centers in AMF_3 hosts, they clearly exhibit a jump of the order parameter at T_c for $RbCaF_3$ (Fig. 1.24), $RbCdF_3$ and $TlCdF_3$.

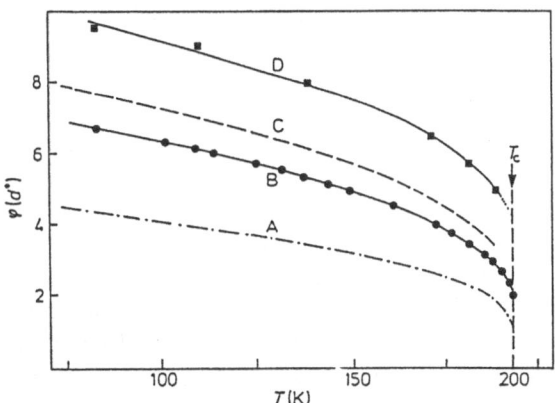

Fig. 1.24. Measurement of the local (EPR) or of the intrinsic order parameter in $RbCaF_3$ (A) through $b_2^0(T)$ for $[GdF_6]^{-3}$ after corrections from theory; (B) Gd^{3+}-O^{2-} pair; (C) neutron scattering [1.170]; (D) through $b_4^4 O_4^4$ and tilts of the corresponding x and y axes for the cubic center. From [1.165a]

The local values of $\varphi(T)$ given by the Gd^{3+}-O^{2-} pair nicely fit the intrinsic value given by neutron scattering [1.170]. The EPR method permits one to measure $\varphi(T)$ directly very near T_c in a temperature range where neutron scattering is inoperative. The local values deduced from b_2^0 for the cubic center (Fig. 1.24) exhibit exactly the same behavior. The quantitative discrepancy arises in particular from an inaccurate theoretical model for converting $b_2^0(T)$ into $\varphi(T)$. The essential point is that all EPR measurements consistently monitor the order-parameter behavior $\varphi(T)$ and its first-order jump.

For all centers, a coexistence of cubic and tetragonal lines is observed in a small temperature interval near T_c at X-band frequencies. The measurements using the IBM K-band variable-temperature cavity enables one to rule out any artefact arising from a temperature gradient or temperature fluctuations. A dynamical model based on two relaxation times has been proposed to account for the temperature dependence of the lines in the X-band spectra near T_c [1.171], but it was not supported by K-band measurements, because the predicted frequency dependence of the spectra was absent.

Briefly, the essential features of a slightly first-order transition are easily deduced from EPR spectra. These were also confirmed by a detailed examination of the shape of the X-ray Bragg line near T_c [1.172]. Therefore, EPR measurements enable us to investigate transitions with slightly first-order character and appear well suited for studying the effects of symmetry-breaking fields, which are theoretically predicted to be of such character.

b) Discontinuous to Continuous Restored SPT Under {001} Uniaxial Stress. In cubic symmetry and no stress applied, mean-field theory predicts a second-order cubic-to-tetragonal transition (Sect. 1.5.2b). However, a small first-order jump of the order parameter can be induced by critical cubic fluctuations, provided they are highly anisotropic, i.e., $f \simeq 1$ in the LGW expression of the free energy (Sect. 1.5.3). Renormalization-group theory also predicts that the transition can be restored to a continuous one on application of a symmetry-breaking tetragonal field, because a near-2D correlated sheet, say (001), is favored over the other two, (100) and (010), all of which are equivalent in cubic symmetry. Actually, the soft R_{25} mode is very flat along the RM line of the Brillouin zone:

$$\alpha = \frac{d^2\omega}{dq^2} = 0.013 = 1 - f$$

in RbCaF$_3$ instead of $\alpha = 0.036$ in SrTiO$_3$. At the R corner $+\varphi - \varphi$ rotations perpendicular to the (001) plane occur, at the M corner $+\varphi + \varphi$ rotations. For a vanishing parameter α, the system would exhibit Lifshitz behavior, as the φ rotations along [001] are undetermined. In this case, the system could reach a tricritical point with a symmetry-breaking field inducing a one-dimensional order parameter [1.100].

Experimentally [1.101] the restoration to a continuous transition has been observed by monitoring the b_2^0 parameter of the cubic center under near [001] stress. [100] and [010] domains are then observed below T_c with a c/a ratio larger than one. The inset of Fig. 1.25 indicates that a continuous transition is restored for $\sigma_t \geq 1.9 \pm 0.3$ kg/mm^2 where the tricritical point is reached at $T_t = 193.3 \pm 0.1$ K. The temperature dependence of $\{b_{2T}^0\}^{2.8} \propto \{\varphi\}^{5.6}$ for the cubic center is represented in Fig. 1.25. The linear dependence of $\{b_{2T}^2\}^{2.8}$ in the second-order regime below $t = (T_c - T)/T = 0.023$ corresponds to a critical exponent $\beta = 0.18 \pm 0.2$. At lower temperatures, the experimental results deviate upwards from the exponential law, at nearly the same reduced temperature and in the same way as observed for monodomain SrTiO$_3$. [100] and [010] domains are present, but always with quite different intensities. A nearly [010] monodomain was obtained for the highest stress allowed by the softness of the crystal. This means that the symmetry-breaking field

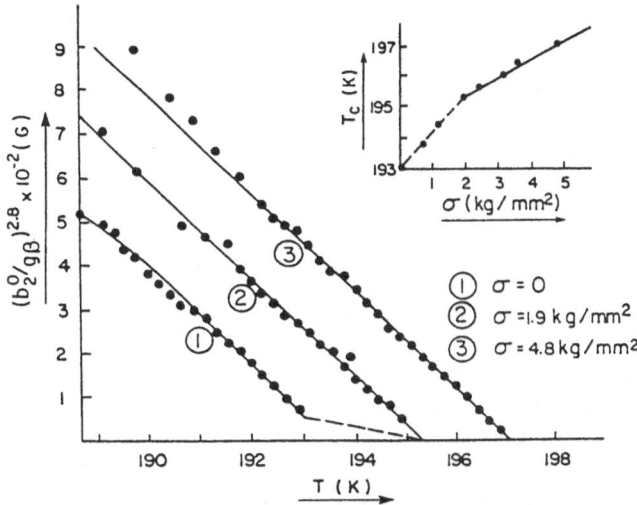

Fig. 1.25. Temperature dependence of $b_2^0(T)$ for [001] stress and [100] domains. The dotted lines outline the dependence of the first-order jump on σ. The inset indicates the dependence of T_c and T_t on σ. From [1.101]

involving both internal and uniaxial stress, with some inhomogeneity, did lead to a system with a one-dimensional order parameter, i.e., the stress was not precisely axial along [001].

A change of the parameters u and v of the LGW equation under hydrostatic stress may also induce tricritical behavior: $u + v > 0 \rightarrow u + v < 0$ [1.173]. Using a K-band hydrostatic-pressure cavity up to 12 kbar, i.e., six times the tricritical stress, the observed linear shift of T_c is given by $dT_c/dP_H = 3.6\,\text{deg/kbar}$ instead of the unaxial shift $dT_c/d\sigma_{(100)} = 12\,\text{deg/kbar}$ $(T > T_t)$. Moreover, no significant change of the first-order jump is observed [1.174], see Fig. 1.26. Therefore, the hydrostatic component of the uniaxial stress $p_\sigma = \frac{1}{3}\sigma_{|100|}$ plays no role, neither in the tricritical behavior nor in the shift of T_c under uniaxial stress. Otherwise one should

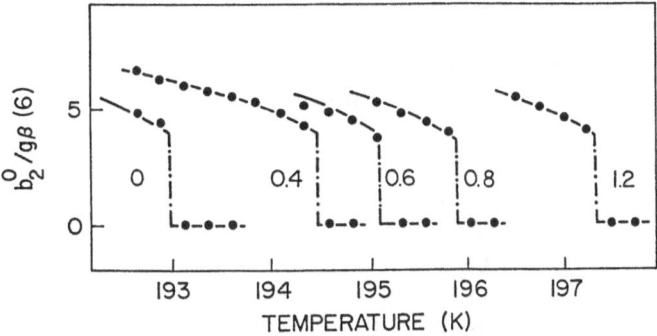

Fig. 1.26. First-order jump and critical temperature dependence on hydrostatic pressure in RbCaF$_3$. From [1.174]

obtain $dT/dp \simeq 36\,$deg/kbar, an order of magnitude larger than the experimental value. Measurements under uniaxial stress and hydrostatic pressure give substantial evidence that the tricritical point is reached through a symmetry-breaking field with $n = 1$. This is a prerequisite for a Lifshitz tricritical point.

The upper limit of β_{exp}, i.e. 0.20, differs considerably from the value $\beta = 0.25$ for normal tricritical behavior. Nevertheless, logarithmic corrections of the form $(t\ln|t|)^{1/4}$, present at low t, could lead to an effective β: $\beta_{\text{eff}} = 0.19$. Such behavior has only been reported for NH_4 or ND_4 halides, which probably exhibit a higher-order critical behavior. The lower limit of $\beta_{\text{exp}} = 0.16$ is consistent with a Lifshitz tricritical behavior, for which the prediction is $\beta = \frac{1}{8}$ to $\frac{1}{6}$. However, a crossover to normal Lifshitz $n = 1$ behavior is not observed for the higher stresses. This may result from too small a uniaxial stress or from insufficient precision in β_{exp}. Indeed, normal Lifshitz behavior could be associated with $\beta = 0.21 \pm 0.03$, according to Monte-Carlo simulations. This would preclude the observation of the crossover from tricritical to normal Lifshiftz behavior.

It turns out that the EPR experiments using the cubic Gd^{3+} center support a Lifshitz behavior rather than a normal one. The center, which is well adapted to the geometry of [001] stress, however, yields data which are not accurate enough for a definite conclusion.

c) A Critical End Point Under $\langle 111 \rangle$ Stress in RbCaF$_3$.

In presence of a discontinuous cubic-tetragonal transition, theoretical investigations by *Kerzberg* and *Mukamel* [1.175] of multicritical points predicted a phase diagram with a critical end point (CEP) under compressional [111] stress: a $q = 3$ state Potts first-order line T_p and an Ising line T_I impinging on it at finite angle and finite stress. Specific-heat measurements on KMnF$_3$ could be qualitatively interpreted in this way [1.176], the nature of the high-temperature transition and the symmetry of the phases not being established. Despite the transition at zero stress being continuous, EPR [1.177] and neutron scattering [1.178] experiments indicated CEP behavior in [111]-stressed SrTiO$_3$, but it was poorly resolved. The EPR experiments on SrTiO$_3$, using the Fe^{3+}-V_O pair, initiated the $\langle 111 \rangle$-stress experiments on RbCaF$_3$, using the Gd^{3+}-O^{2-} pair. The latter were successful [1.179] and are reviewed in the following.

In the cubic phase, the paramagnetic pairs aligned along [100] and [010] are equivalent for $H\|[1\bar{1}0]$ and give a single line centered at 3500 G and K-band. A $\langle 111 \rangle$ stress alone cannot resolve this equivalence, and cannot split this line unless a stress-induced SPT occurs. Let us consider the staggered rotations by $\pm\varphi$ around an axis u_0 of components $((\sqrt{2}/2)\sin\alpha, (\sqrt{2}/2)\sin\alpha, \cos\alpha)$, i.e. $\Delta H_0 = \varphi\cos\alpha$, or around equivalent axes u_1, u_2 through the A_3 $\{111\}$ symmetry element, i.e. $u_1 = (\cos\alpha, (\sqrt{2}/2)\sin\alpha, (\sqrt{2}/2)\sin\alpha)$; $u_2 = ((\sqrt{2}/2)\sin\alpha, \cos\alpha, (\sqrt{2}/2)\sin\alpha)$, Fig. 1.27a. To first order, these tilts would induce a splitting ΔH_i equivalent to staggered rotations around $\{001\}$ by $\varphi_{i(001)} = \varphi([\textbf{001}] \times u_i)$, i.e. $\Delta H_i = \varphi(\sqrt{2}/2)\sin\alpha$, $i = 1, 2$. This is the key to discussing the lines represented in Fig. 1.27 in the same terms as for the $O_h^1 \rightarrow O_{4h}^{18}$ transitions.

For the unstressed sample, T_c was observed at $196.05 \pm 0.10\,$K, and the line splitting given by $\Delta H = B\varphi_{(001)}$ with $B = (125 \pm 1)\,$G/deg. Application of a $\langle 111 \rangle$ stress at low temperature below T_c does not modify the lines corresponding to

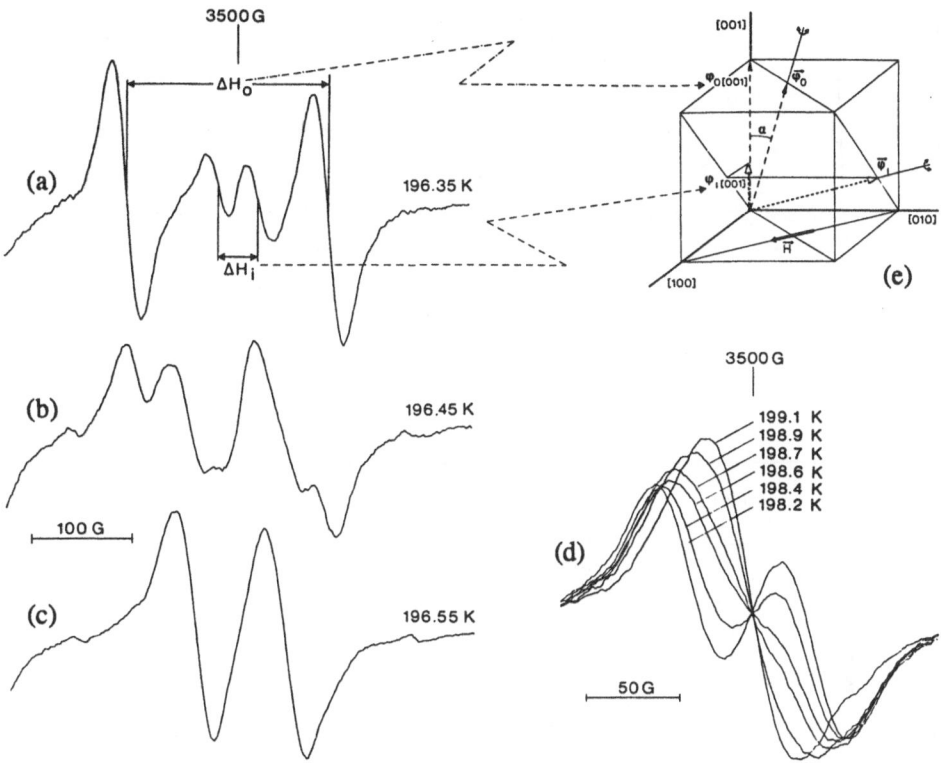

Fig. 1.27a–e. $\sigma_{[111]} = 0.31$ kbar. (a) $T < T_p$: tetragonal phase and line splittings for φ_0 domains (ΔH_0) and φ_i domains ΔH_i; (b) $T \simeq T_p$: coexistence of tetragonal and trigonal phases; (c) $T > T_p$: trigonal phase; (d) second-order transition near T_l; (e) geometry of rotational order parameter vectors and their [001] projection below T_p. From [1.179]

tetragonal {001} domains much. The removal of the tetragonal symmetry is marked by the splitting of the inner lines corresponding to the {100} and {010} domains at zero stress. Without entering into details, the spectrum of Fig. 1.27a ($\sigma = 0.31$ kbar, $T = 196.35$ K) is consistent with an outer splitting ΔH_0 given by rotations around $u_0(\alpha)$ and an inner splitting ΔH_i given by equivalent tilts of and rotations around $u_i(\alpha)$, $i = 1, 2$ in the other domains. Accordingly, α and $\varphi_i[001]$ are given to first order by $\operatorname{tg}\alpha = \sqrt{2}(\Delta H_i/\Delta H_0)$ and by $\varphi_{0,i}[001] = B\,\Delta H_{0,i}$. Warming up the crystal results in an increase of ΔH_i, ΔH_0 being nearly constant. This means an increase of α, i.e., a drift of the u_i towards {111}. It was found that $\alpha = 8.7°$ at 195 K and $16.7°$ at 196.35 K for $\sigma = 0.31$ kbar. At $T_p = 196.45$ K ($\sigma = 0.31$ kbar) the spectrum changes discontinuously to that of Fig. 1.27b. The lines observed below and above T_p coexist. This has to be seen as evidence of a first-order transition. Above T_p, two lines are recorded. They correspond to rotations of the octahedra around the [111] stress axis. Thus, a monodomain sample is obtained with $R\bar{3}c$ structure. When T is raised up to $T_l = 198.95 \pm 0.10$ K, the line splitting, i.e. the order parameter $\varphi_{[111]}$, decreases continuously to zero (Fig. 1.27c,d).

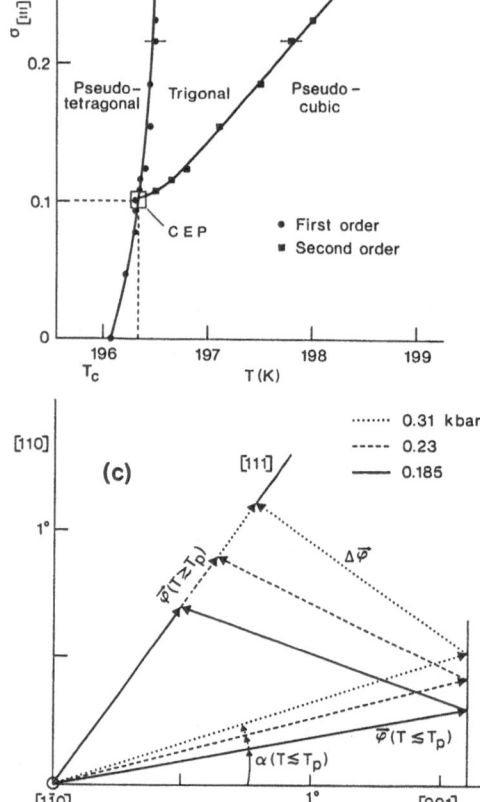

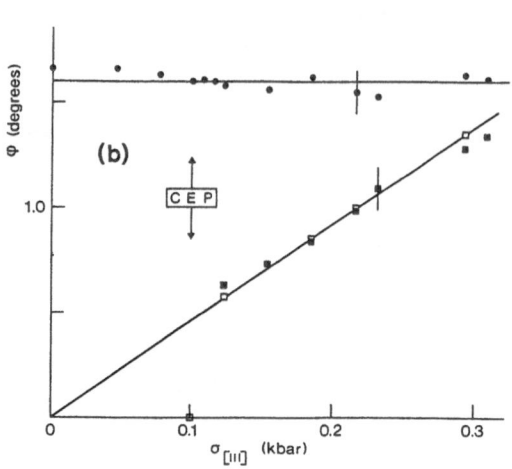

Fig. 1.28. (a) Experimental phase diagram (σ, T) of [111]-stressed $RbCaF_3$. (b) Dots: component φ_0 [0,001] at $T \simeq T_p$ or T_c deduced from $\Delta H_0 = B\varphi_0$; full squares: $\varphi[111]$ at $T \geq T_p$ deduced from line splittings; open squares: $\varphi[111]$ measured by rotating H in the (111) plane. (c) Order parameter evolution at the T_p transition. From [1.179]

The same sequence of phase transitions is obtained for a large range of applied stresses, from $\sigma = 0.31$ kbar down to $\sigma_E = 0.10 \pm 0.1$ kbar. A first-order line separates the pseudo-tetragonal from the trigonal phase (Fig. 1.28a). There T_P and $\varphi_{[001]}$ (Fig. 1.28b) are almost stress independent; φ and α increase with the applied stress, while the order-parameter discontinuity $(\Delta\varphi)$ and the jump of α to 55° at T_P decrease (Fig. 1.28c). From $\sigma = \sigma_E$ down to $\sigma = 0$, the trigonal phase disappears and a first-order transition between a pseudo-tetragonal phase and pseudo-cubic phase [1.179] is observed at T_c. Its essential features do not depend much on the applied stress. The second-order line $T_I(\sigma)$ above σ_E meets the first-order line $T_P(\sigma), T_c(\sigma)$ at a finite angle, see Fig. 1.28a, allowing location of the coordinates of the CEP at $\sigma_E = 0.10 \pm 0.01$ and $T_E = 196.3 \pm 0.1$ K.

Qualitatively, the phase diagram may be easily interpreted in terms of a competition between the "natural" ordering along $\langle 100 \rangle$ and the ordering along [111] favored by the [111] stress. For $\sigma < \sigma_E$, the stress is too small to compete and only perturbs the natural ordering slightly. For $\sigma > \sigma_E$, the stress wins over the natural ordering below T_P. In other words, considering the components of the rotational

54

vector order parameter φ along [111] and in the (111) plane, the first-order transition at T_P corresponds to the ordering of the in-plane components, i.e., to a $q = 3$ state Potts transition [1.175] to relate to the $SrTiO_3$ phase diagram of Fig. 1.21. In that case it remains open whether a CEP occurs near $\sigma_E \simeq 0$ [1.177, 178].

It turns out that the [111]-stress EPR experiments [1.175] corroborate the topology of the predicted phase diagram and its CEP, and allow a determination of the amount and direction of the order parameter in the tetragonal and trigonal phases, see Fig. 1.28b,c, i.e., the experiments proved a quantitative verification of the predictions of the theory.

1.7 Order-Disorder Transitions

A structural transition with a one-dimensional order parameter $n = 1$ implies a double-well local potential for the atomic positions. Pure order-disorder transitions correspond to $kT_c \ll V_m$ (V_m being the depth of the wells) and a very anharmonic potential, see especially the introduction to Vol. I and [1.84]. For displacive transitions, $kT_c \gg V_m$ and the local potential is more harmonic. Upon approaching T_c, the average over the correlated regions becomes more and more anharmonic. A crossover from displacive to order-disorder behavior is expected. It should be marked by precursor-order clusters and by relaxator dynamics. This behavior, favored by a low effective dimensionality and by short-range forces, was initially observed by EPR in $SrTiO_3$ [1.180]. Qualitatively any structural transition should exhibit order-disorder features at sufficiently small $t = (T - T_c)/T_c$ [1.142, 143].

Recently, direct insight into the local potential of $BaTiO_3$ and the collective dynamics of Ti^{4+} ions was obtained by EPR through adapted probes: Fe^{3+} and Mn^{4+}, respectively. Apart from the oxide ferroelectrics (Sect. 1.4), EPR indicates a strong order-disorder character for many ferroelectric transitions and hydrogen-bond ordering [1.181]. In this section, we illustrate a characteristic response of a substitutional EPR probe to an orientational order-disorder transition in NH_4AlF_4, where all essential features (static order, depth of the local wells, precursor-order clusters, collective dynamics, effective lattice dimensionality) can be deduced from the EPR lines [1.182–184]. Moreover, insight from EPR into the mixed crystals $NH_{4(1-x)}Rb_xAlF_4$ is briefly reported, in order to show the possibilities of EPR spectroscopy in problems also studied intensively by NMR spectroscopy (Chap. 2).

In the NH_4AlF_4, (001) layers of $[AlF_6]$ octahedra are connected by the equatorial fluorines and separated by (001) layers of NH_4^+. At room temperature, the $[AlF_6]$ groups are tilted by an angle of $\pm\varphi$ around $c = [001]$. They are antiferro-rotationally ordered along c, as in tetragonal (001) domains of AMF_3 perovskites. The NH_4^+ occupy two equivalent positions which correspond to each other through tilts by $\pi/2$ around [001]. At equilibrium, the H-H edges in the (001) plane are randomly oriented along the a axes. Finally, the displacive Ising-like variable of the system (φ) is ordered at room temperature, whereas the second pseudo-spin type variable involving the NH_4^+ is not (Fig. 1.29).

The average quadratic symmetry at the Al^{3+} site is reflected by a quadrupolar parameter $b_2^0 = 1112 \times 10^{-4}\,cm^{-1}$ for a substitutional Fe^{3+} impurity. In the mixed

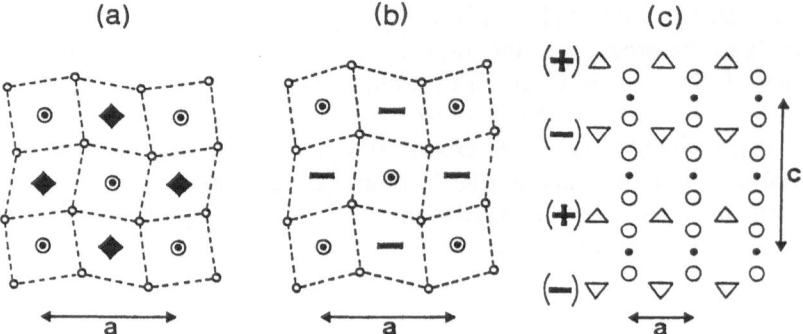

(a)　　　　　(b)　　　　　(c)

Fig. 1.29a–c. Schematic representation of the crystal structure of NH_4AlF_4: (a) disordered phase, projection on (001); (b) low-temperature ordered phase, projection on (001), and (c) order along c. Full circles: Al^{3+} or Fe^{3+} at $z = 0$; small open circles: equatorial fluorines at $z = 0$, large open circles: axial fluorines; squares: disordered NH_4^+ centered at $z = c/2$ (room temperature); bars: H-H edges of ordered NH_4^+. In (c) the triangles stand for the projections of the (NH_4^+) tetrahedra onto (100). From [1.183b]

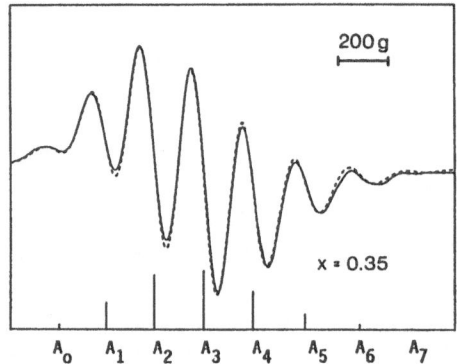

Fig. 1.30. Structure of the $\Delta M_s = \frac{3}{2} \rightarrow \frac{5}{2}$ line and $H\|[001]$ for $Rb_x NH_{4(1-x)}AlF_4$ and $x = 0.35$. A_I represent the lines for $Rb_i NH_{4(1-i)}$ due to the second shell. Full line: experimental; dotted line: reconstruction according to random substitution. From [1.184a]

crystal $Rb_x NH_{4(1-x)}AlF_4$, the lines $\Delta M_s = 1$ for $H\|[001]$ exhibit a well-resolved structure (Fig. 1.30), which respresents the distribution of local crystal fields according to

$$b_2^0(i) = b_2^0(x) + i\Delta b_2^0 \qquad (i = 1, \dots, 8)$$
$$I(i) = x^i (1 - x)^{(8-i)} C_8^i \quad ,$$

where $I(i)$ is the intensity of the ith line.

It could then be inferred that the Fe^{3+} probe mirrors an average crystal field through $b_2^0(x)$, and the *local* organization of the eight second-shell monovalent cations surrounding the probe from a superposition model [1.184]. On the other hand, the probe indicates that no chemical ordering occurs for any x. This allows direct measurement of the concentration x of the random impurity Rb^+ by means of a computer simulation of the structure (Fig. 1.30). Finally, the evolution of $b_2^0(x)$ is such that for $0.1 \leq x \leq 0.25$, the transition $\Delta M_s = -\frac{1}{2} \rightarrow \frac{1}{2}$ for $H\|[001]$ is near a level crossing involving the $M_s = \frac{1}{2}$ and $M_s = \frac{5}{2}$ states.

1.7.1 NH$_4^+$ Ordering in Pure Crystals

The temperature dependence of the high field line ($H = 5700$ G) for $H\|[100]$ is depicted in Fig. 1.31. Qualitatively, it shows a slowing-down stemming from precursor-order clusters represented by an outer doublet below 210 K, and full order below 155 K marked by a static splitting of the low-temperature line. The splitting corresponds to an asymmetry of the local crystal-field with an orthorhombic term of magnitude $b_2^2 = \pm 281 \times 10^{-4}$ cm^{-1}. The equivalent sites have their asymmetry axes tilted by $\pi/2$ around [001]. Therefore the low-temperature phase is also quadratic.

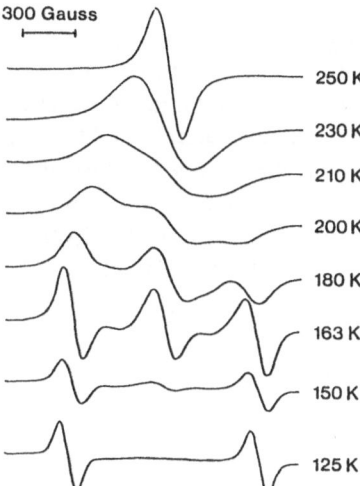

300 Gauss

250 K

230 K

210 K

200 K

180 K

163 K

150 K

125 K

Fig. 1.31. Influence of the slowing down on the line at $H = 5700$ G for $H\|[100]$. From [1.183b]

Assigning the local symmetry breaking to the NH$_4^+$ parallel ordering in the (001) layers and antiparallel ordering (Fig. 1.29) between the layers, the structure could be directly inferred from the EPR spectra. Indeed, any other low-temperature order would lead to a local A$_4$ or $\bar{A}_4$ axis [1.182]. Thus the low-temperature space group was obtained directly from EPR and confirmed by neutron scattering [1.185].

Consistent with the short-range character of the probe as demonstrated by the line structure in mixed crystals, the local configurations of the nearest neighbor NH$_4^+$, located in two adjacent layers, were grouped into four main configurations:

a) Low-temperature local order: the local parameter is $b_2^2 = b_2^2 \, (T < T_c)$.
b) Local parallel order of the nearest NH$_4^+$ in each layer, but parallel order between them. The local symmetry element is $\bar{A}_4$ and $b_2^2 = 0$.
c) Intermediate local order: Parallel order exists between the nearest NH$_4^+$ in one layer, but the nearest NH$_4^+$ in the second layer are disordered. For such configurations $b_2^2 = b_2^2/2 \, (T < T_c)$ was assumed. Indeed, only one half of the NH$_4^+$ cooperate with a well-defined crystal field parameter b_2^2.
d) Local disorder of the nearest NH$_4^+$ in each layer, restoring an average tetragonal symmetry: $b_2^2 = 0$.

The qualitative relevance of this classification is based on an underlying super-position model, apparent in the spectra (Fig. 1.30): configuration (a) (precursor-order cluster) is responsible for the outer doublet, configurations (b) and (d) for the central line, and configuration (c) for an intermediate doublet of "bumps" which becomes apparent in the slow regime below 180 K.

The experimental spectra were fitted to a dynamical model [1.183, 184] involving a relaxation between the configurations, random jumps with many simultaneous reorientations of the NH_4^+ being forbidden. The results are plotted in Fig. 1.32, which depicts the dependence of the occupation probabilities $P(a)$, $P(b)$, $P(c)$, $P(d)$ on T and of the inverse lifetime (τ^{-1}) of precursor-order clusters.

From room-temperature down to 165 K, it was found that $P(a) = P(b)$, Fig. 1.32a. This means that in this temperature range the ordering process is essentially two dimensional (2D): ferro and antiferro orders along c are equivalent. At lower temperature, the onset of a three-dimensional (3D) ordering is marked by a fast decrease of $P(b)$ and a sharp increase in the density of 3D ordered clusters. Defining $\sigma(T)$ as the density of 2D parallel-order clusters in the (001) layers, the occupation probability of 3D configurations $P(c)$ should be given by $2\sigma(1 - \sigma)$ with $P_{c(max)} = \frac{1}{2}$ in a 2D regime. This is experimentally verified. More generally, above $T = 165$ K, any T was characterized by a 2D-ordering parameter σ (Fig. 1.32a) according to

$$P(a) = \tfrac{1}{2}\sigma^2, \quad P(b) = \tfrac{1}{2}\sigma^2, \quad P_c = 2\sigma(1 - \sigma), \quad P(d) = (1 - \sigma)^2 \ .$$

Thus, the dimensionality of the correlations was directly established.

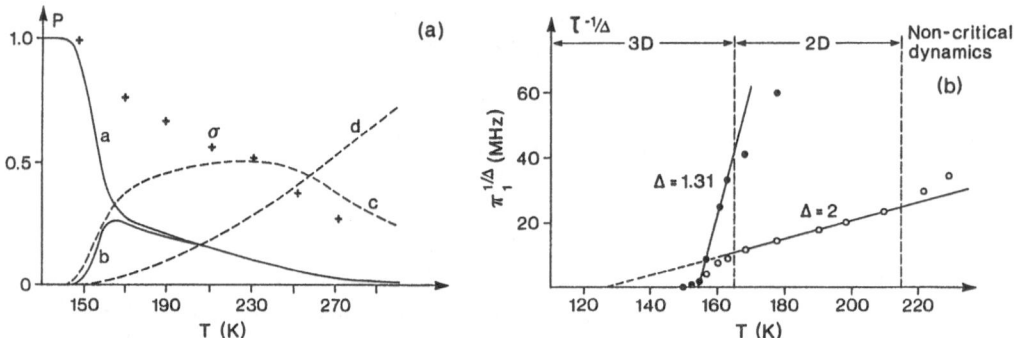

Fig. 1.32. (a) Probabilities of local configurations $P(a)$, $P(b)$: full line; $P(c)$, $P(d)$: dashed line; crosses: 2D ordering parameter σ. (b) Relaxation rate of precursor order cluster. From [1.183b]

According to Fig. 1.32, three different regimes could be observed for τ^{-1}. In the temperature range 170–220 K, τ^{-1} follows a law $\tau^{-1/\Delta} \propto (T - T_{2D})$ with $T_{2D} = 125$ K and $\Delta = 2$. This may evoke a 2D Ising-like slowing-down with $\Delta = Z\nu$, $Z = 2$, $\nu = 1$. At lower temperature τ^{-1} deviates from the 2D regime and follows a law $\tau^{-1/\Delta} \propto (T - T_c)$ with $\Delta = Z\nu$, $Z = 2$, $\nu = 0.65$, i.e., $\Delta = 1.30$, and $T_c = 155$ K. Thus a 2D→3D crossover is apparent in the critical slowing-downs. These regimes may be characterized by critical exponents, proportional to the exponent ν for the correlation length of 2D and 3D Ising systems. However, this result may be fortuitous. Indeed the probe is not well suited for the slow regime below $\tau^{-1} = 10^8$ MHz. A local probe with a longer characteristic time would be required. Above 220 K, τ^{-1} deviates upwards from the 2D regime and follows an Arrhenius law with activation energy $E = 1630$ K, in accordance with the NMR results [1.186]. The NH_4^+ then reorient independently in the local double-well potential $V_M = E = 1630$ K, which is thus determined by the motionally narrowed linewidth in the fast regime of fluctuations.

Near T_c, the nearly static spectrum is marked by a low-temperature doublet and a less intense central quadratic line on a flat absorption background. This characterizes a long-range 2D order, and the quadratic line can be assigned to flat antiphase walls of the low-temperature 3D order. Full low-temperature 3D order is rapidly established below $T_c \simeq 155$ K.

1.7.2 NH_4^+ Ordering in Mixed Crystals: $Rb_xNH_{4(1-x)}AlF_4$

Up to Rb concentrations of $x = 10\%$, the critical slowing-down qualitatively exhibits the same features as in Fig. 1.31. Each line is marked by satellites due to the nearest neighbor Rb^+, and becomes smeared out for $H\|[100]$ and $x > 10\%$.

The low-temperature lines for $H\|[100]$ are represented in Fig. 1.33 for $x = 0.03$ and 0.07. An outer doublet, as in the low-temperature phase of pure crystals, and a central quadratic line corresponding to a second shell of eight NH_4^+ are observed. These lines are perturbed by one, two, or more Rb^+ ions randomly substituted for the NH_4^+. The computer simulation indicates that the Rb^+ ions randomly perturb the doublet and the quadratic line. Thus the quadratic sites are not preferentially pinned by the Rb^+. On the other hand, no absorption background corresponding to disorder is apparent.

In keeping with the above, the central $\Delta M_s = -\frac{1}{2} \rightarrow \frac{1}{2}$ transition yields a central quadratic line for $H\|[001]$, and a doublet (Fig. 1.34) which arises from an accidental spin-level mixing of $M_s = \frac{1}{2}$ and $M_s = \frac{5}{2}$ states by a substantial b_2^4 term as for the low-temperature phase in pure crystals; see inset of Fig. 1.34. The quadratic line

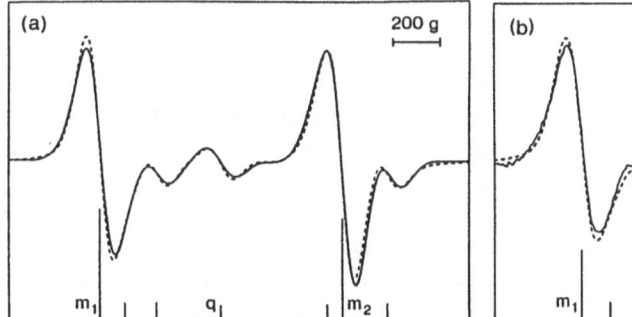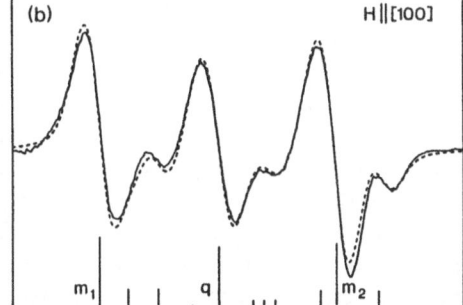

Fig. 1.33a,b. Structure of the line centered at $H = 5700$ G for $H\|[100]$ in mixed $Rb_xNH_{4(1-x)}AlF_4$ crystals: (a) 4% and (b) 6.7% of Rb. The doublet (m_1, m_2) and the central line q represent local "antiferro" and "ferro" order, respectively. The small vertical lines represent sites perturbed by 1, 2 or 3 Rb^+ ions consistent with random substitution. From [1.184a]

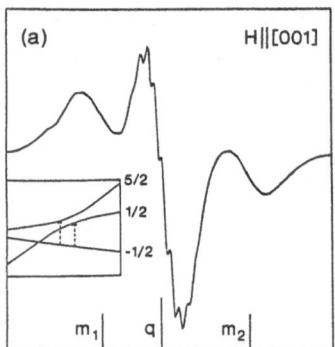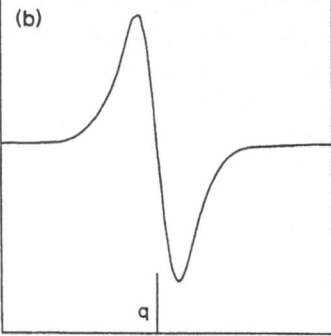

Fig. 1.34a, b. Structure of the line $\Delta M_s = -\frac{1}{2} \to \frac{1}{2}$ for $H\|[001]$: **(a)** 6.7%, **(b)** 25% of Rb. The inset indicates the mechanism of the level crossing for local "antiferro" order. From [1.184a]

up to $x = 10\%$ exhibits a well-resolved SHF ^{19}F structure, not resolved at room temperature, which marks a well-defined static local order. For this orientation, the perturbations by the Rb^+ ions are not resolved. All these observations are consistent with a rather long-range 2D parallel order of the NH_4^+ in the (001) layers. Doping the crystal with Rb^+ destroys a full antiparallel order between adjacent layers.

Adjacent layers with parallel order, which may be antiphase walls, are represented by quadratic lines which persist down to very low temperature. For $x < x_{c_1} = 4\%$, the density of antiphase walls is given approximately by x. The order of pure crystals is not essentially modified. In agreement, recent neutron investigations indicate that the superstructure Bragg lines of the low-temperature phase are present [1.184].

For $x \geq x_{c_1}$, the intensity of the quadratic line increases drastically: a new phase occurs. Possible high-order commensurate phases, such as given by extended Ising models with competing interactions [1.187], could account for the spectra. Neutron scattering does not give evidence for such phases. The actual phase for $x > x_{c_1}$ is therefore similar to a pseudo-spin glass, with long-range 2D ordered domains and 3D disorder.

For $x > 10\%$, the lines for $H\|[100]$ become hard to analyze. The evolution of the low-temperature phase can be followed by the $\Delta M_s = -\frac{1}{2} \to \frac{1}{2}$ line for $H\|[001]$, see Fig. 1.34. For $x = 25\%$, a quadratic line is observed. The SHF interaction is no longer resolved, owing to line broadening through topological disorder. The NH_4^+ exhibit essentially a 3D parallel order. The critical concentration at which the ferro order occurs has not yet been determined.

The phase diagram (x, T) of the $NH_{4(1-x)}Rb_xF_4$ system turns out to exhibit the essential features of a universal behavior. The freezing of the collective reorientations of the NH_4^+ occurs near $T = 150\,\mathrm{K}$ to produce antiferro order, which does not vary much for small x. For $x \simeq x_{c_1} \simeq 4\%$, a steep line separates the antiferro order from the pseudo-spin glass. For $x = x_{c_2}$ ($13\% < x_{c_2} < 25\%$), another transition line separates the glassy phase from the ferro order. The same phase diagram was observed for $NH_{4(1-x)}Rb_xH_2PO_4$ and for magnetic alloys with competing ferro-antiferro interactions [1.188].

1.7.3 Room-Temperature Phases of Mixed Crystals

In Sect. 1.7.2, the ordering of the NH_4^+ sublattice was considered to be independent of the organization of the $[AlF_6]$ sublattice. In pure NH_4AlF_4, the staggered tilts of the $[AlF_6]$ octahedra in the (001) layers are antiferro-rotationally ordered along c at room temperature. In pure $RbAlF_4$, they are ferro-rotationally ordered. For both crystals, $|\varphi_{001}| \simeq 9°$. In mixed crystals, therefore, competing interactions involving the Ising-like variable φ_{001} are likely to be present.

Figure 1.35 gives the values of the local crystal field parameter $b_2^0(i)$ (i defining the local second shell i Rb^+-$(8-i)NH_4^+$) as a function of concentration x of $RbAlF_4$. For a critical concentration x_c of about 40% and for different samples, two values of $b_2^0(i)$, i.e., two types of crystals, were observed simultaneously and depending on the batches of hydrothermal growth, i.e., on the exact conditions (P, T) of the growth. For $x > x_c$, $b_2^0(i)$ does not depend significantly on x, and the fluorine octahedra are probably ferro-rotationally ordered along c as in pure $RbAlF_4$.

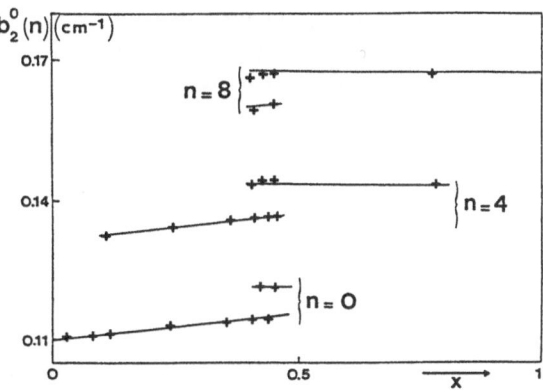

Fig. 1.35. Local field parameter $b_2^0(i)$ for $Rb_iNH_{(8-i)}$ at room temperature for $i = 0, 4, 8$. From [1.184a]

For $x < x_c$, the crystal-field parameter evolves linearly towards the value of pure $RbAlF_4$. One may therefore question whether the order of the fluorine octahedra and the amplitude of the displacive order parameter $\{\varphi_{001}\}$ along c are influenced by the orientational ordering of the NH_4^+. The Fe^{3+} probe is not local enough to measure the local tilts of the octahedra, and thus is not sufficiently suited to monitor the rotational order between adjacent layers of $[AlF_6]$ groups. For this last purpose, a probe substituted for the monovalent ion, i.e., between two adjacent layers of AlF_6, would be more appropriate. Let us recall here that in $SrTiO_3$ the antiferro distortive order of the low-temperature phase could be fully demonstrated by using two probes, one substituted for Ti^{4+}, the other for Sr^{2+}. At present, we cannot exclude a disorder in the organization of the $[AlF_6]$ layers, which ultimately drives the freezing of the NH_4^+ for $0 < x < 25\%$. This is still an open question.

Summarizing, we have described the various ways in which the EPR method can be used to investigate pure order-disorder transitions and transitions in topologically disordered systems. Its merits and achievements are quite clear: the discovery of the existence of a transition and the identification of the low-temperature order, a direct observation of precursive order clusters and information about their dynamics, direct information about the anisotropy of the correlations and of their interactions with very simple experiments. Nevertheless, the time scale of the Fe^{3+} probe is too short to monitor the slow-motion regime near T_c accurately.

The Fe^{3+} probe allows one to monitor the configurations of blocks of pseudo-spins via three types of lines. One may consider here the first step of a Kadanoff transformation for the NH_4^+ lattice. Blocks into which the Rb^+, diamagnetic-like impurity enters can also be probed. The range of the probe is such that one can only detect relatively long-range NH_4^+ 2D order in the glassy phase, and cannot monitor the tilts of the fluorine octahedra. Moreover, the probe location is not well suited to monitor the 3D order of the fluorine octahedra. This illustrates the limits of the method when a single type of probe is used.

A review of electron spin and paramagnetic resonance in KH_2PO_4 and its iso-morphs has recently appeared in [1.189]. Its abstract reads: "This paper reviews magnetic resonance of radicals and paramagnetic ions in KH_2PO_4-type crystals up to 1985. In two sections, dedicated to each category of impurities, their site, symmetry and static particularities are summarized. Then the application of spin resonance to monitor fast ferroelectric domain polarization reversal is exposed. Finally, in the last section, slow reorientations of certain magnetic defects as well as information on intrinsic soft-mode dynamics from a static defect are described."

1.8 Incommensurate Phases

Before reviewing EPR in structural incommensurate (INC) phases, we want to re-call the basic features of magnetic resonance lines in such phases. More details can be found in Chap. 2 and in [1.190, 191]. Usually one observes on cooling at higher temperature a commensurate-incommensurate transition T_I and at lower T an incommensurate-commensurate transition T_C.

i) A static one-dimensional (1D) modulation in the plane wave regime of an INC phase below T_I, $u_M = A_T u_0 \cos(\Phi_M + \Phi_0)$, results in a shift of the EPR line [1.192] at point M:

$$\Delta H(\Phi_M) = A h_1 \cos(\Phi_M + \Phi_1) + A^2 \{ h_2' + h_2'' \cos 2(\Phi_M + \Phi_2) \} \quad . \qquad (1.32)$$

$A \propto (T_I - T)^\beta$ represents the critical amplitude, $\beta = 0.35$ the critical (XY-like) exponent, and $\Phi_M = q_I \cdot 0M$ the local phase with q_I the INC wave vector and $0M$ the local radius vector.

The expression for $\Delta H(\Phi_M)$ is a power series expansion up to second order in the local order parameter $A e^{j\Phi_M}$. The parameters (h_i, Φ_i) depend on the microscopic details of the spin lattice interaction. The phase parameters Φ_1, Φ_2 arise from first- and second-order contributions of atomic displacements with different phases within the range of the probe (consistent with the wave concept). They may also arise from a coupling in quadrature of two displacement modes [1.192, 193]. Either the first-order or the second-order terms can be forbidden by symmetry for particular orientations of the magnetic field. Setting $\Phi_1 = \Phi_2 = \Phi_0$ and $h'_2 = h''_2$ leads to the "local phase approximation", which was used initially [1.190].

In the plane wave regime, all the values of the phase are equally probable. It turns out that the line shape should then be given by

$$F(H - H_0) = \int_0^{2\pi} f\left(\frac{H - H_0 + \Delta H_\Phi}{L}\right) d\Phi \quad . \tag{1.33}$$

H_0 represents the resonance field in the parent commensurate (COM) phase; f and L stand for the shape and width of the local line at the resonance field $H_0 + \Delta H_\Phi$.

The line shape is marked by singularities corresponding to $\partial \Delta H / \partial \Phi = 0$. Two edge singularities are obtained for either first-order or second-order dominant contributions. Generally three singularities are obtained for equivalent contributions. Only a phase difference $(\Phi_1 - \Phi_2)$ may result in a fourth singularity. At a lower temperature $T < T_C$, the INC wave vector locks at a commensurate value q_C. Above and near T_C, the "multi-soliton" regime takes over from the plane-wave regime. In the limit of narrow solitons, a regular array of domains of the low-temperature phase are separated by discommensurations. Consequently, discrete EPR lines of the low-temperature phase superimposed on a continuous absorption background should be observed.

ii) Competing interactions drive the evolution of q_I to q_C which may pass through high-order commensurate values and, correspondingly, to stable or metastable higher-order commensurate phases. This is the devil's staircase, which may be complete (no stable INC phase) or incomplete [1.194]. Universal models, derived from the Ising model, taking into account competing n.n., n.n.n., and n.n.n.n. interactions have recently been developed to deal with this particular aspect [1.187, 195].

iii) The incommensurate phase may exhibit two vectors with different modulation components along different crystal axes x and y. For the star of wave vectors having a dimension $n = 4$, it may not be easy to distinguish a 2D modulation (q_i^1 and q_i^2) from a 1D modulation in equivalent domains (q_i^1 or q_i^2) by scattering techniques [1.196].

Here, our purpose is to report simple examples illustrating the above three possibilities (i–iii) by EPR spectroscopy with various probes, and to give a more general bibliography of the problem.

Furthermore, within a harmonic approximation, incommensurate phases are described by particular phonons as amplitude and phason modes. Roughly speaking, the amplitude mode near T_i has a normal soft-mode behavior. In contrast, the phason mode has, in principle, a zero gap at $q = q_I$, and corresponds to a modulation of the local crystal field. Few direct observations of this mode are known. However, we may infer large spectral densities of phase fluctuations within the time scale of EPR spectroscopy, persisting far below T_i. They may give rise to a line broadening which depends on the local phase Φ. This is actually observed, and is detailed below.

In Sect. 1.8.1 on ThBr$_4$ and ThCl$_4$ and Sect. 1.8.3 on diphenyl, we shall deal with pure plane wave regimes and with INC phases where the phason mode has been directly observed. The probes are, respectively, the Gd^{3+}-$V_{[x-]}$ defect, comparable to the Fe^{3+}-$V_{O^{2-}}$ pair in $SrTiO_3$, and the excited triplet state of guest molecules with two or three phenyl rings. With BCCD (betaine calcium chloride dihydrate) in Sect. 1.8.2, we shall restrict ourselves to a mainly visual examination of spectra showing a devil's staircase behavior and a multi-soliton regime. The probe is a substitutional impurity. We shall give further references to EPR work using such probes.

Currently, several EPR investigations of CDW's (charge density waves) are in progress. The EPR lines of paramagnetic impurities are affected by the CDW order parameter and by the application of dc voltage near the threshold voltage [1.197]. The mechanism of the spin-CDW interaction remains to be clarified. For instance, the paramagnetic impurity may polarize the CDW locally, allowing it to interact with the spin probe [1.198]. The CDW may also modulate the crystal field. A review of this evolving field is probably premature.

1.8.1 EPR of Charge-Compensated Defects in the INC Phases of ThBr$_4$ and ThCl$_4$:Gd^{3+}

ThBr$_4$ is probably the best example of a pure displacive COM$\rightarrow$INC phase transition in ionic crystals where all the basic features (the soft phonon above T_I, the amplitudon and the phason below T_I, as well as the static 1D plane wave modulation) really correspond to direct observations by scattering methods [1.199, 200]. As usual, the scattering techniques may suffer from a lack of accuracy and resolution near the critical temperature. This crystal and its isotype ThCl$_4$ will be used to exemplify the response of an EPR probe to an INC phase and the particular information available very near T_I.

In these crystals, the Th^{4+} are surrounded by flat tetrahedra of n.n. halide ions and by elongated tetrahedra of n.n.n. halide ions (Fig. 1.36). The COM$\rightarrow$INC transition which occurs at $T_I = 95$ and $70\,\mathrm{K}$ for ThBr$_4$ and ThCl$_4$, respectively, results in a modulated displacement field $[q_I = (\frac{1}{3} - \delta)c^*]$, consisting of tilts of the tetrahedral ligands around the c-axis. A local binary axis is preserved for the Th^{4+} site alone. The displacement mode arises from the coupling in quadrature of two modes. They

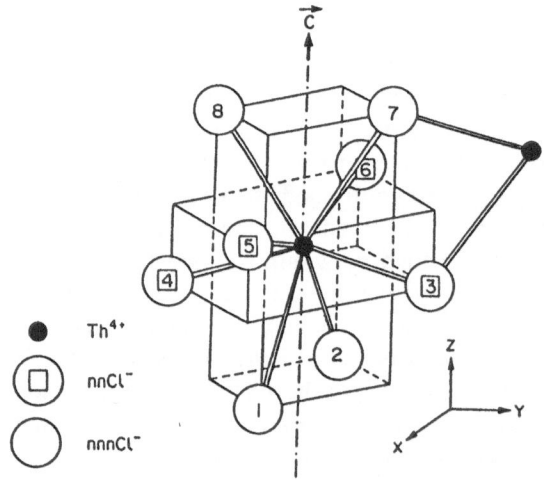

Fig. 1.36. Unit cell of ThBr$_4$ and ThCl$_4$. The [100] axis is along 3–4. In ThCl$_4$:Gd^{3+} the Cl$^-$ vacancy is located at sites 3–4–5–6. In ThBr$_4$:Gd^{3+} the Br$^-$ vacancy is located at sites 1–2–7–8. Reprinted with permission from [1.201]. Copyright 1986 Pergamon Journals Ltd.

correspond, for $q = 0$, to a global rotation and a twist of the surrounding ligand, respectively [1.199].

In crystals doped with Gd^{3+}, charge compensation occurs principally through the pairs Gd^{3+}-$V_{n.n.n.Br-}$ and Gd^{3+}-$V_{n.n.Cl-}$, where $V_{n.n.n.Br-}$ for instance stands for a next-nearest neighbor Br^- vacancy. They were identified with EPR. Indeed, the spin Hamiltonian in the commensurate room-temperature phase could be analyzed in terms of a quadratic Hamiltonian due to the host, perturbed by an axial Hamiltonian due to the vacancy whose direction nicely fits the Th-n.n.n. Br and the Th-n.n. Cl directions in the host [1.192, 201].

Despite the local symmetry breaking by the defect, the symmetry of the modulation wave only allows second-order line shifts for $H\|[100]$ (Fig. 1.36), while both first-order and second-order line shifts are allowed for $\vartheta = ([100], H) \neq 0$ [1.201, 202]. The experimental results are depicted in Fig. 1.37 for the $\Delta M \simeq \frac{5}{2} \rightarrow \frac{7}{2}$ transition and $T_I - T = 10$ K. Introducing a phase-dependent linewidth $L(\Phi) = L_0 + L_1 \cos^2(\Phi + \tilde{\Phi}_1)$, a quantitative agreement between theory and experiment was found. Qualitatively, the rotation of H in the (001) plane away from $\langle 100 \rangle$ induces an increasing first-order contribution, and the number of singularities

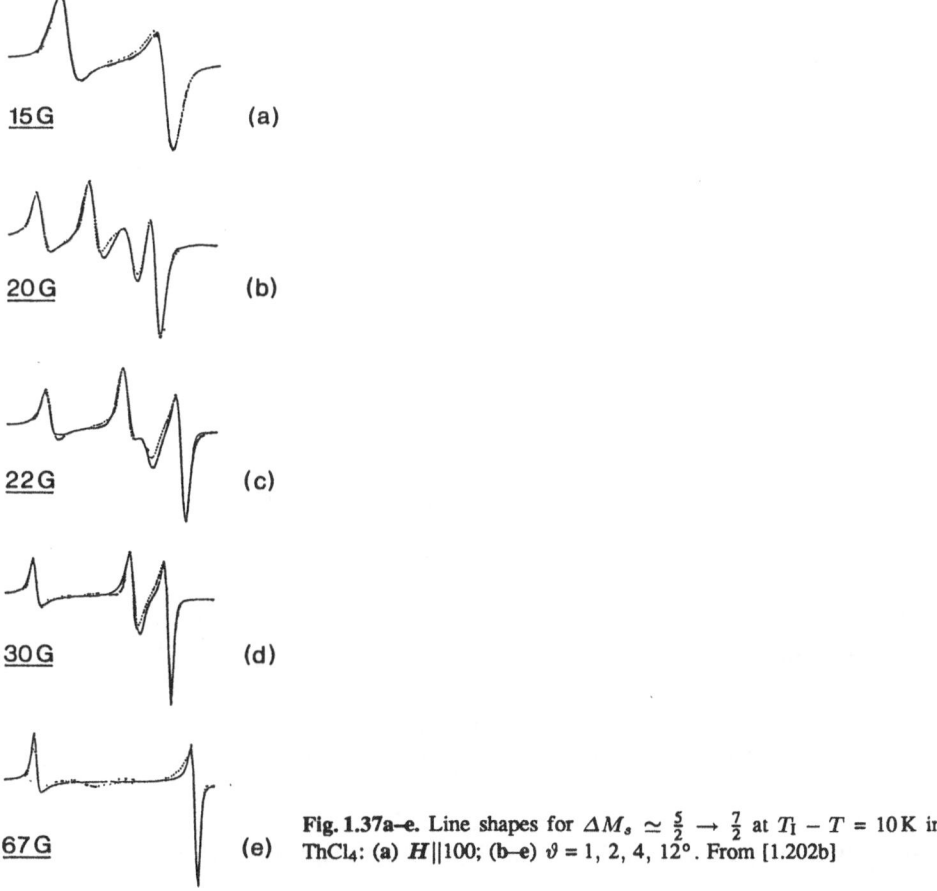

15 G   (a)

20 G   (b)

22 G   (c)

30 G   (d)

67 G   (e)

Fig. 1.37a–e. Line shapes for $\Delta M_s \simeq \frac{5}{2} \rightarrow \frac{7}{2}$ at $T_I - T = 10$ K in ThCl$_4$: (a) $H\|100$; (b–e) $\vartheta = 1, 2, 4, 12°$. From [1.202b]

found verifies the number of solutions for $\partial H/\partial \Phi = 0$. Four solutions occur for $\vartheta = 1°$, $2°$ through a phase difference $\Phi_1 - \Phi_2 \simeq 30°$. For $\vartheta = 12°$, the singularity splitting is about 400 G at X-band frequencies. Such a high sensitivity reminds one of the Fe^{3+}-V_O pair in $SrTiO_3$. Both probes are well suited to monitor tilts of metal ligands of the host. Moreover, the excellent agreement between theory and experiment confirms a pure plane wave regime and precludes any phase pinning by the defect. On the other hand, a partial phase pinning by the substitutional U^{4+} impurity in $ThBr_4$ was observed in luminescence spectra at 4 K [1.200]. This may illustrate the importance of the electronic configuration of an extrinsic probe, as previously discussed. Note that Fig. 1.37 depicts the whole set of possible line shapes in a 1D plane-wave regime.

Similar results were obtained in $ThBr_4$ but in a different way [1.192]. The ratio of first-order to second-order contributions varies as $A^{-1} \propto (T_{\rm I} - T)^{-\beta}$ for orientations of the magnetic field allowing first- and second-order contributions. Consistent with this, a sequence of 2,3,4 singularities was observed on cooling. In both crystals, the amplitude was found to vary critically with $\beta = 0.35$ in a large temperature range below $T_{\rm I}$. The local amplitude of the tilts estimated from the splittings and from the superposition model [1.192, 202] was in satisfactory agreement with the results of neutron scattering [1.199].

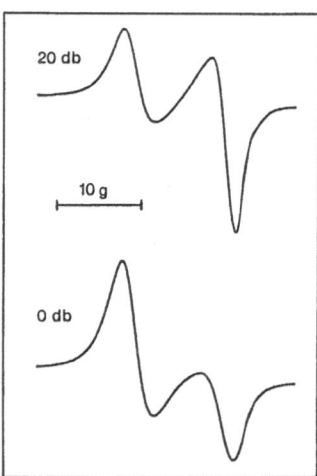

Fig. 1.38. Line shape at $T_{\rm I} - T = 15$ K for $\Delta M_s = \frac{1}{2} \to \frac{3}{2}$ and $H \| 100$, showing saturation effects. From [1.202b]

For $H \| [100]$, the two edge singularities were found to be significantly asymmetric for some transitions [1.192, 202], for instance the $\Delta M_s \simeq \frac{1}{2} \to \frac{3}{2}$ line in $ThCl_4$ (Fig. 1.38). The sharp high-field singularity can easily be saturated, as seen in the figure, implying a phase-dependent T_1. This was considered as a sign of large, persisting phase fluctuations in the frequency range of 10^{10} Hertz far below $T_{\rm I}$, according to the following possible scheme. A dominant first-order Hamiltonian $A\mathcal{H}_1 \cos \Phi$ was assumed to be responsible for the static line shift via second-order perturbation. The sharp singularity should then correspond to $\Phi \simeq (0, \pi)$ and the broad one to $\Phi \simeq (0, \pm \pi/2)$. For the latter, it can be assumed that no effect of underdamped amplitude fluctuations is present, because of a large gap and the underdamping of the amplitude mode [1.200]. Then the "classical-lattice" fluctuating spin Hamiltonian should be $A\mathcal{H}_1 \sin \Phi \delta \varphi(t)$, implying a short relaxation time for $\Phi = \pi/2$, i.e. for the broad singularity. This qualitative interpretation is appealing for true measurements of $T_1(\Phi)$ with the INC EPR lines, as in NMR spectroscopy but with a time scale modified by two orders of magnitude. Recent developments in EPR technology make them more quantitative.

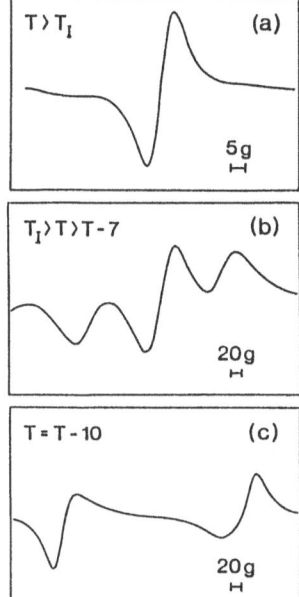

Fig. 1.39a–c. Coexistence of "INC-like" and "COM-like" lines near T_1 in ThCl$_4$: (a) and (c) pure commensurate and incommensurate lines, $\{\vartheta = 8°\}$. In (b), both a commensurate-like line and edge singularities exist. Reprinted with permission from [1.201]. Copyright 1986 Pergamon Journals Ltd.

Near T_1, both ThCl$_4$ and ThBr$_4$ exhibit a coexistence of "INC edge singularities" and of a central "COM-like" line [1.192, 202] in a temperature range of a few K, see Fig. 1.39. For the orientation employed, the first-order line shift is dominant. A similar spectrum was observed for ^{87}Rb-NMR in Rb$_2$ZnBr$_4$ [1.203], and attributed to a motional averaging of the INC modulation by large phase fluctuations in defect-free regions of the crystal and to a pinned modulation near defects. Full pinning caused by the critical increase of the amplitude arises a few K below T_1 [1.203].

For ThCl$_4$, a precise location of the critical temperature was established via a change in the behavior of the "COM-like" line. The Lorentzian (l) and the Gaussian (g) parts of the linewidth, and the shift of the center of the line ΔH are plotted in Fig. 1.40 versus temperature. A simultaneous and sharp change of all parameters is observed at $T = T_1 = 70\,\mathrm{K}$. Moreover, just below T_1 and keeping a stationary temperature, the width of the essentially Lorentzian shape was shown to depend quadratically on the linear static line shift for angles ϑ ranging from 3° to 15° [1.202]. According to an argument previously used for SrTiO$_3$ [1.204], this is consistent with a motional averaging accounting for the "COM-like" line below T_1.

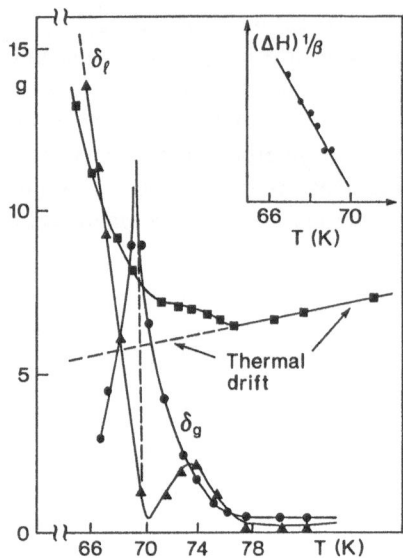

Fig. 1.40. Temperature dependence of the parameters of the "COM-like" line near T_I: squares: center of line; circles: Gaussian width δ_g. The "λ-shaped" variation evokes a normal behavior for line widths at a SPT; triangles: Lorentzian width δ_l. Inset: Plot of $\{\Delta H\} 1/\beta$ versus $T (\beta = 0.35)$; ΔH represents the distance between the edge singularities. From [1.202b]

1.8.2 EPR of Mn^{2+} in Doped Betaine Calcium Chloride Dihydrate (BCCD) and Substitutional Probes

At room temperature, the space group of BCCD is $Pnma$. The Mn^{2+} substitutes for Ca^{2+} at crystal sites which exhibit local (010) mirror-plane symmetry [1.205]. The EPR investigations [1.205] were initiated following previous X-ray and dielectric measurements [1.206, 207].

For $H\|b$, all sites are equivalent and a single low-field hyperfine sextuplet was observed at room temperature (Fig. 1.41). The evolution of this sextuplet on cooling down to 100 K is represented in the sequence of spectra in Fig. 1.41. At 164 K, a continuous change leads to INC spectra, in which each hyperfine line gives rise to two edge singularities. At 127 K, a change into a discrete set of lines is observed. A computer reconstruction of one hyperfine line, Fig. 1.42a, indicates that seven discrete lines are present, an indication of a new unit cell seven times larger than that at room temperature. From 124 K down to 116 K, a two edge singularity spectrum is first restored, then modified by a superstructure of increasing intensity corresponding to a narrow-soliton regime, which leads to a sharp change at 116 K.

At this temperature, down to 73 K, each initial line is split into a discrete quadruplet which may reflect a unit cell enlarged by a factor of 4. At 73 K, a discontinuous change occurs. Despite an overlap between the hyperfine lines, an accurate reconstruction of their structure yields a set of five lines with a small, but relevant unresolved splitting, see Fig. 1.42b, i.e., a $q = \frac{1}{5}$ commensurate structure. Another smooth change occurs at about 47 K. At 15 K, a discontinuous change restores a hyperfine sextuplet as observed at room temperature, see Fig. 1.43. Therefore the size of the RT unit cell is restored ($q = 0$) but with a lower symmetry as indicated by satellite lines corresponding to $\Delta M_I \neq 0$.

The incomplete devil's staircase previously observed by X-ray Bragg scattering [1.206] with a sequence [COM($q = 0$) →INC→COM($q = \frac{2}{7}$) →INC →COM

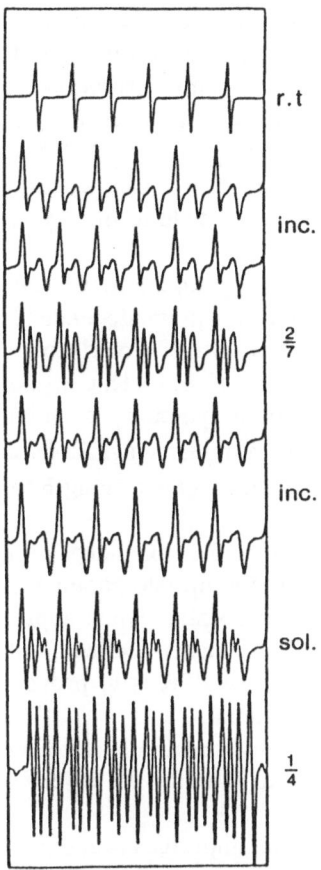

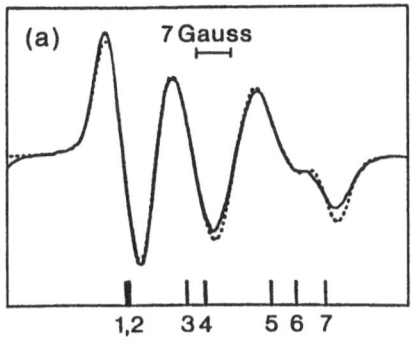

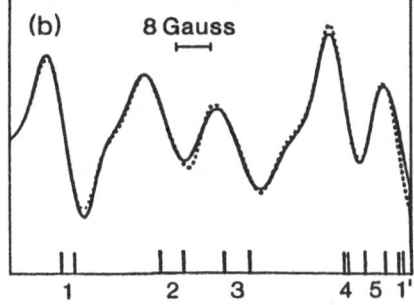

Fig. 1.42. Detail of a hyperfine line and computer reconstruction showing seven components (**a**) and five doublets (**b**), respectively, for $q = \frac{2}{7}$ and $q = \frac{1}{5}$. (1′) in (**b**) belongs to the adjacent hyperfine transition. *Solid line*: experiment, *dotted line*: reconstruction. From [1.205]

Fig. 1.41. Temperature dependence of the low-field hyperfine sextuplet ($\Delta M_I = 0$, $-\frac{5}{2} \leq M_I \leq \frac{5}{2}$) for $\boldsymbol{H} \| \boldsymbol{b}$. Note the rise of a narrow soliton regime in the second INC phase. From [1.205]

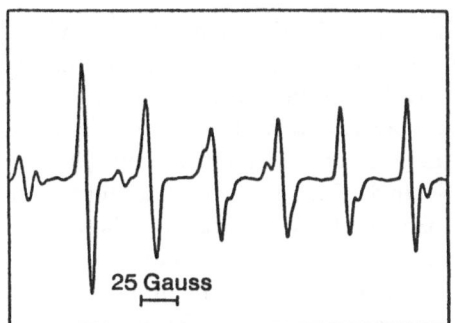

Fig. 1.43. Hyperfine sextuplet below $T = 15\,\text{K}$. The symmetry lowering with respect to room temperature (Fig. 1.41) is marked by $\Delta M_I \neq 0$ satellites. From [1.205]

$(q = \frac{1}{4}) \rightarrow COM(q = \frac{1}{5}) \rightarrow COM(q = \frac{1}{6})$; $q = qc^*$] is confirmed by the EPR method with the help of an easy experiment. Different regimes of the modulation in the INC phases are depicted qualitatively, the values of the higher-order commensurate wave vectors appear clearly, except for $q = \frac{1}{6}$, and apparently a new low-temperature phase with $q = 0$ occurs at $T = 15$ K in BCCD:Mn^{2+}, not observed in pure crystals. More information on the symmetry of the high-order commensurate phases can probably be obtained by considering different orientations of the magnetic field to the crystal axes.

For $H\|b$ and $q = \frac{2}{7}$, the positions of the split lines (Fig. 1.42a) fit the law $H(n) = H_0 + h \cos^2(2\pi nq + \Phi_0)$; $n = 1, \ldots, 7$, consistent with a quadratic dependence on the local amplitude of the order parameter. The same law, which would give a doublet instead of the observed quadruplet for low temperatures (Fig. 1.41), does not apply for $q_c = \frac{1}{4}$. This may suggest the occurrence of an extra symmetry breaking for $q = \frac{1}{4}$. Actually, the most valuable information about the symmetry available so far is given by the measurements of the spontaneous polarization, either along b for $q = \frac{2}{7}$ or along a for $q = \frac{1}{6}$ [1.207].

In searching for some universal behavior of the system within extended Ising or related models involving competing interactions [1.194], appropriate parameters could account for the full sequence of the incomplete "devil's stairs" when appropriate trajectories in the phase space are chosen. This was for instance realized with the (TMA)$_2$MCl$_4$ compounds [1.195]. The particular interest of EPR is to prove a phase with $q_c = 0$, i.e., to evidence the ultimate step of the stairs.

The interesting case of BCCD certainly deserves further investigations. Here, we have illustrated the efficiency of simple EPR experiments to monitor the essential features of a complex system.

Many other INC phases have been studied by substitutional impurities or radicals. In K$_2$(SeO$_4$) an incommensurate phase occurs at 129.5 K and was investigated with the help of the (SeO$_4^-$) radical obtained by γ-ray irradiation [1.208]. The details of the line shape at 110 K were interpreted in terms of a broad soliton regime [1.208]. In Rb$_2$ZnCl$_4$, NQR on ^{35,37}Cl ($I = \frac{3}{2}$), NMR on ^{87}Rb ($I = \frac{3}{2}$) and EPR on Mn^{2+} ($S = \frac{5}{2}$, $I = \frac{5}{2}$) substitutional for Zn^2 ($I = 0$) allow all crystal sites to be monitored by local measurements. Despite a nice fit with the host and the good sensitivity of the EPR spectrum to the structural transitions $\{$Pnma, $z = 4\} \xrightarrow{T_I = 30^\circ C}$ INC $\xrightarrow{T_c = -78^\circ C}$ COM ($q_c = a^*/3$), it is not so easy to obtain quantitative information from the probe. Indeed the hyperfine structure results in an overlap of the lines which are rather broad because of an underlying disorder. There are two ways to resolve this: First, the hyperfine interaction allows a very particular and accurate method to monitor the static critical behavior near T_I [1.209, 210]. Second, since the hyperfine interaction is a nearly constant atomic property of the probe, the EPR spectrum can be cleared by signal treatments [1.211, 212]. Let us begin with these technical points.

Above T_I, the [MnCl$_4$]$^{2-}$ groups possess a (010) mirror. The main local axis of the quadrupolar spin-lattice interaction lies in this plane with $(a, z^\pm) = \pm 7^\circ$ for the two magnetically inequivalent sites. $x = b$ is a common quadrupolar axis. For H in the (a, b) plane, all sites become magnetically equivalent by symmetry. On the other hand, nuclear spin transitions $\Delta M_s = \pm 1$ are forbidden for H along the local

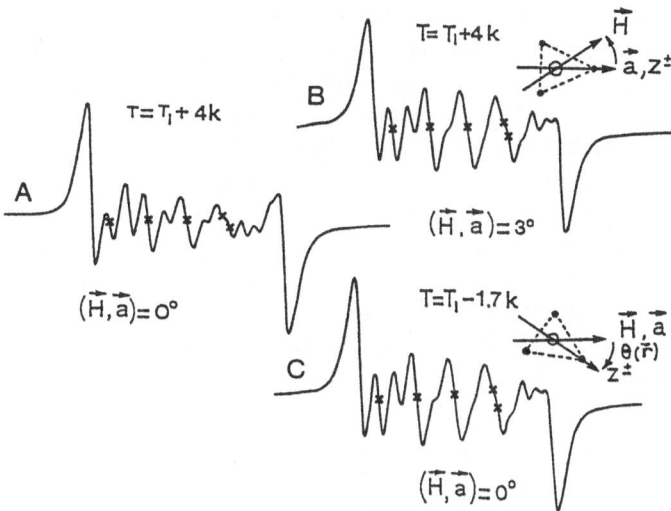

Fig. 1.44a–c. Hyperfine structure of $\Delta M_s = -\frac{1}{2} \to \frac{1}{2}$. (a) $T > T_{\mathrm{I}}$ and $H\|a$; (b) $T > T_{\mathrm{I}}$ and H tilted, and (c) $T < T_{\mathrm{I}}$ and $H\|a$. The clearly resolved lines $\Delta M_{\mathrm{I}} = \pm 1$ are indicated by a star

quadrupolar axes (x, y, z). Otherwise they become allowed and can reach a large probability for $\Delta M_s = -\frac{1}{2} \to \frac{1}{2}$. They are responsible for the complicated hyperfine structure for $H\|a$ (Fig. 1.44a). Their intensity increases sharply, see Fig. 1.44b, when H is slightly tilted by θ_c around c, according to

$$\Delta I(\theta_c) = \left\{ I(\theta_c) - I(H\|a) \right\} \propto \theta_c^2 \quad .$$

Below T_{I}, in the INC phase, the nearly rigid$[MnCl_4]^{2-}$ groups undergo a modulated tilt around a and around c. For the sake of simplicity, we may ignore associated distortions and assume that essentially the probe is sensitive to the configuration of the first shell of ligands according to the superposition model, see Sect. 1.2.4. Then the EPR spectrum is not much influenced by the tilts around a, and the local tilts around c by $\theta(r) = \theta_0 \cos \phi(r) = \theta_0 \cos(kr + \phi_0)$ are equivalent to a local tilt by $-\theta(r)$ of the magnetic field, see Figs. 1.44b and c. Therefore the average intensity of the hyperfine lines $\Delta M_{\mathrm{I}} = \pm 1$ increases for $H\|a$ according to

$$\Delta I(T < T_{\mathrm{I}}) = \left\{ I(T) - I(T_{\mathrm{I}} + \varepsilon) \right\} \propto \langle \theta_0^2 \cos^2 \phi(r) \rangle = \frac{\theta_0^2}{2} \propto [T_{\mathrm{I}} - T]^{2\beta} \quad .$$

Let us note that the essential result, i.e., the quadratic dependence of $\Delta I(T < T_{\mathrm{I}})$ on the amplitude of the order parameter, is not due to the shortcomings of the tilt model. It only depends on the symmetry, i.e., on the local breaking of the (010) plane by the plane wave modulation near T_{I}.

Experimentally the hyperfine structure (Fig. 1.44a) does not change down to a well defined temperature T_{I}. Then a sharp and continuous modification occurs, but the line shape remains commensurate-like. This can be accurately verified by the nice reconstruction of the structure with a discrete set of lines, see Fig. 1.45.

Moreover, the smallest details of the hyperfine structure below T_{I}, for any T in the range $(T_{\mathrm{I}}, T_{\mathrm{I}} - 6\,\mathrm{K})$, can be experimentally reproduced above T_{I}, by a convenient tilt $\theta_c(T)$ of H (Figs. 1.44b and c). This means that $\theta_c(T)$ is an accurate image of the amplitude of the order parameter at the temperature T. Therefore monitoring the critical behavior of the order parameter only requires a visual examination of two sets of lines: those recorded for $H\|a$ and T varying smoothly below T_I, and those recorded for $T = T_I + \varepsilon$, H becoming smoothly tilted by θ_c. Fortuitous reasons favor this rather odd way of monitoring the static critical behavior. In any case, it is certainly one of the simplest and probably one of the most accurate for an incommensurate system. Usually dynamical effects and lack of a perfect long-range order very close to T_I make the observation of the true static critical behavior difficult. Alternatively, the reconstruction results (Fig. 1.45) can be used to monitor the critical amplitude through $\Delta I(T)$. The (x, y)-like behavior predicted by the theory ($n = 2$) is well verified, see Fig. 1.46.

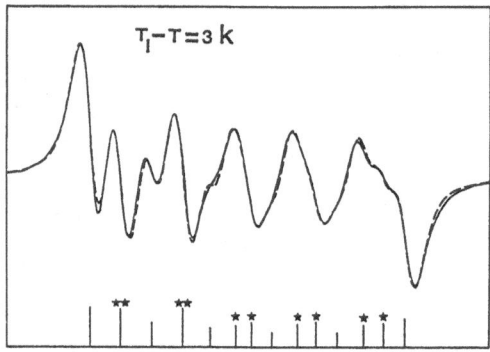

Fig. 1.45. Computer reconstruction of the structure below T_I with discrete commensurate-like lines. ($\Delta M_I \pm 1$ transitions are indicated by a star over the bars)

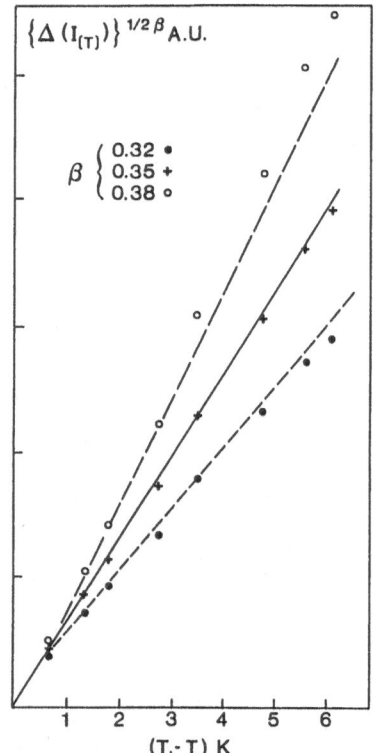

Fig. 1.46. Plot of $\{\Delta I(T)\}^{1/2\beta}$ versus $(T_I - T)$. A ▶ linear law is obtained for $\beta = 0.35$

On the other hand, dealing with the lines $\Delta M_s = \pm\frac{3}{2} \leftrightarrows \pm\frac{5}{2}$ is much more complicated. Nevertheless for this transition each local, fine-structure line splits into a constant sextuplet ($\Delta M_I = 0$) not influenced much by the transitions $\Delta M_I = \pm 1, \pm 2$. Therefore it was possible to deconvolute the experimental lines below T_I by the hyperfine structure recorded for $T > T_I$. This numerical treatment [1.211, 212] clears out the essential features of the modulation. Particularly, the rise of the multisoliton regime near T_c is marked by the rise of "embryos" of the lines which are fully developed below T_c in the COM phase, see Fig. 1.47. The rise of the multisoliton

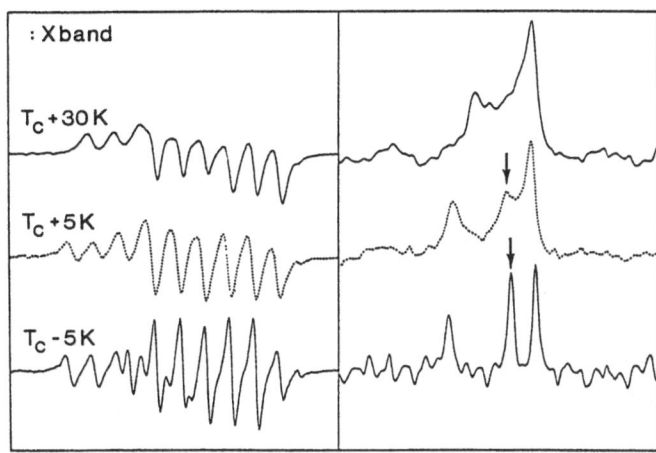

Fig. 1.47. The embryo of a low-temperature line at $T = T_c + 5\,\mathrm{K}$ becomes apparent when the experimental line is deconvoluted by the hyperfine structure. This line is marked by an arrow, and becomes fully developed below T_c. From [1.212]

regime, also marked by slow kinetic phenomena, was directly observed early on [1.213]. The numerical treatment now allows a quantitative insight.

For $H \| a$, the local line shift depends quadratically on the amplitude of the modulation. Assuming that the phase is driven by a sine-Gordon equation, allowing a slight modulation of the amplitude, the experimental line shapes near T_c could be reconstructed accurately by a convolution procedure involving the theoretical phase-dependent line and the constant hyperfine structure, and by matching the soliton density. The results are shown in Fig. 1.48 [1.212].

An interesting result [1.212] was obtained for samples cleaved from Bridgman batches and submitted to several thermal treatments prior to the measurements. For

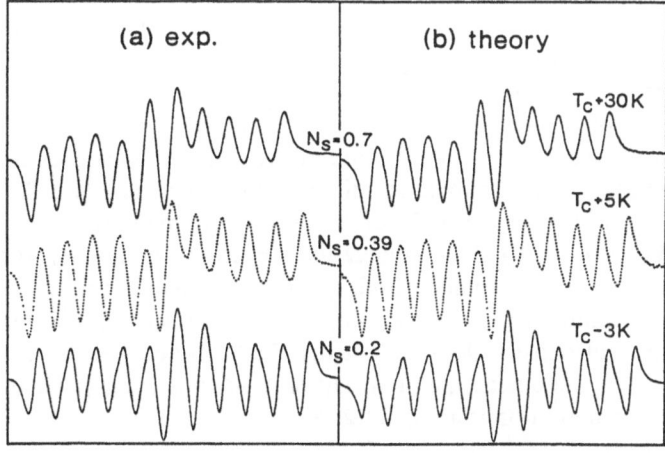

Fig. 1.48. Reconstruction of the experimental line according to a convolution procedure: {theory}* {hyperfine structure}. N_s represents the fitted value of the soliton density. From [1.212]

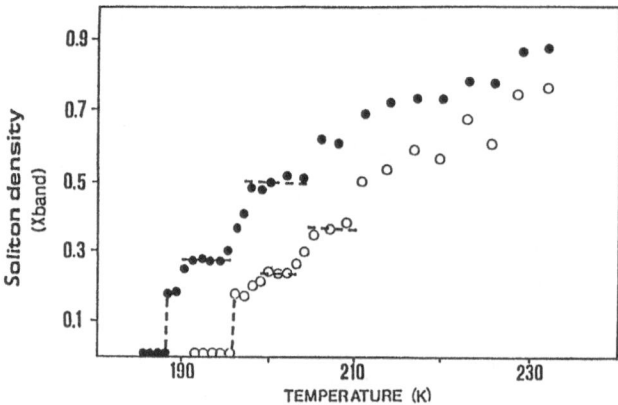

Fig. 1.49. Temperature dependence of the soliton density for a slow thermal cycling. From [1.212]

a very slow thermal cycling by steps of 0.2 K with thermal stabilization between the steps, metastable states were observed over a temperature range of a few K, see Fig. 1.49, [1.212]. They are characterized by a stationary soliton density. Metastable states have also been observed [1.214] in $\{N(CH_3)_4\}_2ZnCl_4$ subjected to large doses of X-ray irradiation inducing defects. The metastable states were characterized by stationary Bragg satellites, which imply some long-range order between the pinned solitons. On the contrary, the local measurements in Rb_2ZnCl_4 do not yield any direct information on the long-range order. Therefore a full identification of the metastable states near the lock-in transition requires the use of both techniques.

The large hysteresis seen in Fig. 1.49 and the dielectric measurements, which show a smearing of the dielectric peaks at T_c, indicate that the behavior of the Bridgman samples near T_c is dominated by defects. It is worth noting that freshly cleaved crystals subjected to a fast thermal cycling (steps of 1 K) do not exhibit metastable states [1.212].

Dynamical phenomena have also been observed using the EPR Mn^{2+} probe and the $\Delta M_s = \pm\frac{3}{2} \leftrightarrows \pm\frac{5}{2}$ lines. The critical slowing down is marked by a central peak phenomenon, in agreement with the results of NMR investigations. This could be inferred from a drastic effect of a small tilt by θ_c of H on the line width. Indeed, the tilt of H away from a allows a linear contribution of the ordering variables on the secular part of the fluctuating Hamiltonian via direct secular mechanisms. The drastic broadening of the line width very near T_I for H slightly oriented away from a requires a large spectral density $J_1(0)$ of the fluctuation spectrum. In contrast, for $H\|a$, the secular part of the spin Hamiltonian depends quadratically on the amplitude of the ordering variables. No significant line broadening is observed near T_I and this means that the relevant spectral density $J_2(0)$ is not drastically influenced by the slowing down. "Central peak" phenomena and a strong order-disorder character of the transition may account for these observations [1.215].

Below T_I and near T_I, large phase fluctuations inducing a partial motional averaging of the line positions were observed [1.211]. This effect prevents a reliable measurement of β near T_I through the amplitude of the local line shifts and enforces the interest of observing the intensities of $\Delta M_I \pm 1$ transitions within $\Delta M_s = -\frac{1}{2} \to \frac{1}{2}$, as described above.

In the plane regime at lower temperature, the asymmetry of the "edge singularity" spectrum (Fig. 1.49) has been assigned to persistent phase fluctuations with large spectral densities in the range of 10^{10} Hz and to local line broadenings through nonsecular mechanisms, i.e., through direct relaxation processes. The same effect has been reported for $ThCl_4$, Fig. 1.38. Using a theoretical model involving an over-damped underlying phason mode, an upper limit for the phason gap could be estimated: $\omega_p \simeq 10^{10}$ Hz [1.211].

Further information about the use of the Mn^{2+} probe in Rb_2ZnCl_4 can be found in [1.216]. Using unusual ways, i.e. transition probabilities, or more sophisticated methodology, i.e. rather heavy numerical treatments, the investigations on Rb_2ZnCl_4 show that EPR probes are also efficient for complicated and messy spectra. Surveying the investigations of Rb_2ZnCl_4 with Mn^{2+} yields that EPR remains efficient despite confusing spectra due to overlap or an intrinsic complexity of the system. Nevertheless, the usual basic quality of the method, i.e., fast access to essential features, becomes spoiled by sophistication of the techniques.

In a tri-sarcosine chloride crystal, an EPR investigation with Mn^{2+} probe ions showed anomalous line broadening above $T = 10°C$, correlated with a change of regime by a temperature dependence of most parameters of the spin Hamiltonian, and also correlated with a distribution of local strains. The line shape resulting from two equivalent sites but discernible for H in the $(a\,c)$ crystal plane was qualitatively interpreted in terms of an underlying, INC-modulated disorder [1.217]. At the ferroelectric transition ($T_c = 130$ K), the coexistence of "paraelectric-like" and "ferroelectric" lines was interpreted in terms of an incommensurate wave resulting from the collective fluctuations of a dipolar pseudo-spin [1.181] and giving rise to similar phenomena as for $ThCl_4$ near T_I. Qualitatively, components of ferroelectric, order-disorder, and COM→INC transitions were invoked to account for the critical behavior of the EPR lines at $T_c = 130$ K [1.181].

1.8.3 Incommensurate Phases in Diphenyl Studied via a Triplet-State Spin Probe

In the molecular crystal diphenyl, $(C_{12}H_{10})$, guest molecules of phenanthrene $(C_{14}H_{10})$ or of naphthalene $(C_{10}H_8)$ exhibit a long-lived triplet state under UV irradiation. The corresponding EPR lines undergo a splitting at 42 K and a further modification at 15 K [1.218]. This was attributed to a structural phase change, initially suggested by Raman investigations. This was actually the first evidence of a COM→INC transition in molecular crystals, although this conclusion was not stated explicitly.

The technique used was "EQR", i.e., electron quadrupolar resonance perturbed by the Zeeman effect, rather than conventional EPR spectroscopy. Furthermore, carbene $[-\dot{C}-]$ has a triplet ground state, which was used to investigate INC phases in partially polymerized crystals [1.219].

Below $T_I = 42$ K, pure diphenyl is known [1.196] to exhibit an INC phase (II) with $q_I(II) = \pm \delta a^* \pm (1 - \delta_b)b^*/2$, and below $T_2' = 17$ K, another INC phase (III) with $q_I(III) = (1 - \delta_b)b^*/2$. In the high-temperature phase (I) the crystal structure is monoclinic with space group $P21/a (Z = 2)$. In the INC phases, the displacement field is essentially a modulated twist of the molecules around their long axis [1.196]. It is worth noting that two components of $q_I(II)$ are incommensurate and that the star of equivalent wave vectors has a dimension $n = 4$. The II→III transition appears as a partial lock-in. The current problem concerning phase II is that neutron scattering can hardly distinguish a 1D modulation $q_I^l = \delta a^* + (1 - \delta_b)b^*/2$ or $q_I^{ll} = \delta a^* - (1 - \delta_b)b^*/2$ in a bidomain structure from a 2D modulation with q_I^l and q_I^{ll} in a single domain. Both modulations are possible solutions of an $n = 4$ LGW equation [1.220].

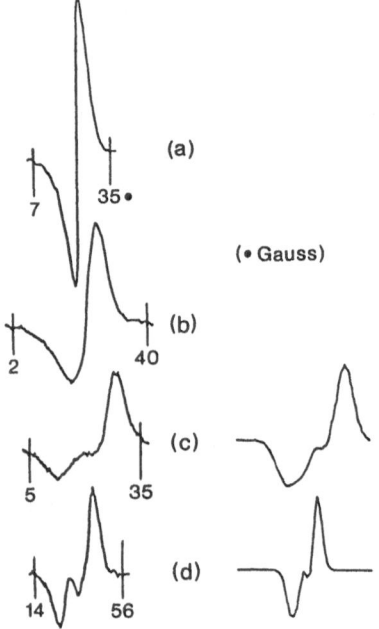

Fig. 1.50. Experimental lines in diphenyl near T_I for (a–d) $T = 42.3, 41.5, 40.5$ and 37.1 K, respectively. From [1.218]. For (c) and (d) computer reconstructions with a 1D model involving first- and second-order line shifts are also given. From [1.221]

Let us re-examine the early EPR results. The experimental line shapes near T_I in Fig. 1.50 are drawn from [1.218] and [1.221]. They can be reconstructed by a 1D plane-wave model. On decreasing the temperature, the distribution of crystal-field parameters $[b_2^0 - b_2^2]$ and b_2^2 progressively exhibits three and four peaks [1.218], i.e., a similar behavior to the crystal-field parameters of Gd^{3+}-V_{Br} in $ThBr_4$. On the other hand, a computer reconstruction of line shapes for a 2D plane wave modulation on the same principles as for a 1D modulation can hardly account for the experimental results [1.221]. This brief discussion of line shapes from early experimental results, which focussed on the triplet state and the conformation of molecules with phenyl rings, supports 1D rather than 2D modulation, but the line shape should be examined in a larger temperature range. It outlines an interesting possibility of the EPR method: the direct application of local measurements to infer reciprocal space properties. Only a few 2D independent modulations are well identified. For instance, they occur for potassium strontium niobate ($n = 4$) [1.222] and for the charge density waves in the 2H-TaSe$_2$ structure ($n = 6$).

In summary, EPR or ESR spectroscopy is an efficient method for obtaining valuable information about COM→INC phase transitions with simple experiments. For the various extrinsic probes employed, such as charge-compensated defects (ThBr$_4$, ThCl$_4$), substitutional paramagnetic impurities (Rb$_2$ZnCl$_4$, BCCD), free radicals (K$_2$SeO$_4$), molecular triplet states, there is no evidence of drastic phase-pinning effects by the probe, i.e., an easy slide of the modulation wave is possible. If this were not the case, it would indeed have considerably reduced the value of the EPR method for investigating the static and dynamic properties of incommensurate phases.

Acknowledgements. The circumstances in which the first author was involved in the past three years were such that without the constant encouragement and insistence of Professor A. Rigamonti as well as his help with the pre-edited version, this chapter would not have come into existence. Our thanks go Dr. Th. von Waldkirch, with whom this article was initiated and who wrote Sect. 1.3.1 but was unable to participate further, owing to his demanding administrative tasks at the ETH Zurich. While finalizing this chapter, we often thought of Walter Berlinger, who upon completion of the original manuscript carefully checked the literature cited, renumbered it and the figure captions, in part, before his untimely and unexpected death. Many of the experiments reviewed here were made possible by his great expertise in EPR. During the preparation of the manuscript, a number of secretaries have contributed with great skill and dedication. At IBM's publications department, these were first Pat Theus and then Charlotte Bolliger, whose contribution in producing this final version, with expert galley-proof reading, is herewith gratefully acknowledged.

References

1.1   K.A. Müller: In *Structural Phase Transitions and Soft Modes*, Proceedings of the NATO Advanced Study Institute, Geilo, 1971, ed. by E.J. Samuelson, E. Andersen, J. Feder (Universitetsforlaget, Oslo, Bergen, Tromsö 1971)

1.2   K.A. Müller, Th. v. Waldkirch: In *Local Properties at Phase Transitions*, Proceedings of the Enrico Fermi Intl. School of Physics Course LIX, Varenna, 1973, ed. by K.A. Müller, A. Rigamonti (North-Holland, Amsterdam 1976) p. 187

1.3   G.E. Pake, T.L. Estle: In *Frontiers in Physics. The Physical Principles of Electron Paramagnetic Resonance*, 2nd Ed., ed. by D. Pines (W.A. Benjamin, Reading MA 1973)

1.4   A. Abragam, B. Bleaney: *Electron Paramagnetic Resonance of Transition Ions* (Clarendon, Oxford 1970)

1.5   S.A. Altshuler, B.M. Kozyrev: *Electron Paramagnetic Resonance* (Academic, New York 1964)

1.6   N.M. Atherton: *Electron Spin Resonance* (Wiley, New York 1973)

1.7   E.J. Zavoisky: J. Phys. (USSR) **9**, 211 (1945)

1.8   M. Tinkham: In: *Group Theory and Quantum Mechanics* (McGraw-Hill, New York 1964) p. 326

1.9   J. Gaillard, P. Gloux, K.A. Müller: Phys. Rev. Lett. **38**, 1216 (1977);
       K.A. Müller, W. Berlinger: Z. Phys. B **31**, 151 (1978)

1.10   D.J. Newman, W. Urban: Adv. Phys. **24**, 793 (1975)

1.11   C.P. Slichter: In: *The Principles of Magnetic Resonance* (Harper & Row, New York 1953)

1.12   E.S. Kirkpatrick, K.A. Müller, R.S. Rubins: Phys. Rev. **135**, A86 (1964);
       Th. von Waldkirch, K.A. Müller, W. Berlinger: Phys. Rev. B **5**, 4324 (1972)

1.13   J.Y. Buzaré, J.J. Rousseau, J.C. Fayet: J. Phys. (Paris) **38**, L-445 (1977)

1.14   K.A. Müller, W. Berlinger: J. Phys. C **16**, 6861 (1983)

1.15   K.A. Müller: Promotionsarbeit ETH Zürich, Prom. Nr. 2791, in: Helv. Phys. Acta **31**, 173 (1958)

1.16   C. Kittel, M. Luttinger: MIT Technical Report, No. 49 (Sept. 10, 1947)

1.17   J. Michoulier, J.M. Gaite: J. Chem. Phys. **56**, 5205 (1972);
       J.M. Gaite, J. Michoulier: J. Chem. Phys. **59**, 488 (1973);
       J.M. Gaite, J. Michoulier: In *Proceedings of the 15th Ampère Congress*, Turkey, 1972, ed. by V. Hovi (North-Holland, Amsterdam 1973) p.207;
       J.M. Gaite: J. Phys. C**8**, 3887 (1975)
1.18   R.R. Sharma, T.P. Das, R. Orbach: Phys. Rev. **149**, 257 (1966); Phys. Rev. **155**, 338 (1967)
1.19   A. Leblé: "Determination et interpretation des paramètres de l'Hamiltonian de spin des ions ^6S dans des cristaux fluoriés" Thesis, University of Le Mans (June 1982)
1.20   J.R. Gabriel, D.F. Johnston, M.J.D. Powell: Proc. Roy. Soc. London Ser. A**264**, 503 (1961)
1.21   S. Geschwind: Phys. Rev. **121**, 363 (1961)
1.22   B. Derighetti, J.E. Drumheller, F. Laves, K.A. Müller, F. Waldner: Acta Cryst. **18**, 557 (1965)
1.23   W. Low, E.L. Offenbacher: In *Solid State Physics*. Advances in Research and Applications, Vol. 17, ed. by F. Seitz and D. Turnbull (Academic, New York 1965) p.136
1.24   S. Geschwind: In *Hyperfine Interactions*, ed. by A.J. Freeman, R. Frankel (Academic, New York 1967)
1.25   G.W. Ludwig, H.H. Woodbury: Phys. Rev. Lett. **7**, 240 (1961)
1.26   W.B. Mims: *The Linear Electric Field Effect in Paramagnetic Resonance* (Clarendon, Oxford 1976)
1.27   M. Weger, E. Feher: In *Paramagnetic Resonance*, Proceedings of 1st Intl. Conference held in Jerusalem, July 16–20, 1962, Vol. II, ed. by W. Low (Academic, New York 1963) p.628
1.28   S.H. Wemple: Thesis, Massachusetts Inst. of Technology, Cambridge, MA (1963)
1.29   A.M. Stoneham: Rev. Mod. Phys. **41**, 82 (1969) and references therein;
       G. Amoretti, C. Fava, V. Varacca: Z. Naturforsch. **37a**, 536 (1982)
1.30   G.E. Stedman, D.J. Newman: J. Phys. C**7**, 2347 (1974);
       J.Y. Buzaré, M. Fayet-Bonnel, J.C. Fayet: J. Phys. C**14**, 67 (1981)
1.31   E.J. Bijvank, H.W. den Hartog, J. Andriessen: Phys. Rev. B**16**, 1008 (1977) and references therein
1.32   N.R. Lewis, S.K. Misra: Phys. Rev. B**25**, 5421 (1982)
1.33   M. Heming, G. Lehmann: Chem. Phys. Lett. **80**, 235 (1981)
1.34   A. Leble, J.J. Rousseau, J.C. Fayet, C. Jacoboni: Solid State Commun. **43**, 773 (1982)
1.35   H.S. Murietta, J.O. Rubio, M.G. Aguilar, J. García-Solé: J. Phys. C**16**, 1945 (1983)
1.36   D.J. Newman, E. Siegel: J. Phys. C**9**, 4285 (1976)
1.37   E. Siegel, K.A. Müller: Phys. Rev. B**19**, 109 (1979)
1.38   Y. Akishige, T. Kubota, K. Ohi: J. Phys. Soc. Jpn. **50**, 3964 (1981)
1.39   M.J.L. Sangster: J. Phys. C**14**, 2889 (1981)
1.40   D.J. Newman: J. Phys. C**15**, 6627 (1982)
1.41   J.F. Clare, S.D. Devine: J. Phys. C**16**, 4415 (1983)
1.42   Y.Y. Yeung, D.J. Newman: Phys. Rev. B**34**, 2258 (1986)
1.43   C.P. Poole, Jr.: In *Electron Spin Resonance. A Comprehensive Treatise on Experimental Techniques* 2nd Ed. (Wiley, New York 1983)
1.44   W. Berlinger, K.A. Müller: Rev. Sci. Instrum. **48**, 1161 (1977)
1.45   W. Berlinger: Rev. Sci. Instrum. **53**, 338 (1982); Magnetic Resonance Review **10**, 45 (1985)
1.46   K.A. Müller: In *Magnetic Resonance in Condensed Matter – Recent Developments*, Proceedings of the IVth Ampère Intl. Summer School, Pula, 1976, ed. by R. Blinc, G. Lahajnar (University of Ljubljana, Yugoslavia 1977), p.637
1.47   C.J. Gorter: Physica **3**, 995 (1936)
1.48   J. Schmidt, I. Solomon: J. Appl. Phys. **37**, 3719 (1966)
1.49   W.S. Moore, T.A. Al-Sharbati: J. Phys. D**6**, 367 (1973) and references therein
1.50   F.I.B. Williams: Proc. Phys. Soc. **91**, 111 (1967)
1.51   H. Bill: Solid State Commun. **17**, 1209 (1975)
1.52   P. Wysling, K.A. Müller: J. Phys. C**9**, 635 (1976)
1.53   J.C.M. Henning, J.H. den Boef: Phys. Rev. B**14**, 26 (1976); Phys. Rev. B**18**, 60 (1978)
1.54   J.H. den Boef, J.C.M. Henning: Rev. Sci. Instrum. **45**, 1199 (1974)
1.55   F. Jona, G. Shirane: *Ferroelectric Crystals* (Pergamon, London 1962)
1.56   A.W. Hornig, R.C. Rempel, H.E. Weaver: J. Phys. Chem. Solids **10**, 1 (1959)
1.57   L. Rimai, G.A. deMars: Phys. Rev. **127**, 702 (1962)
1.58   S.H. Wemple: MIT Technical Report 4251 (1962), p.53; Bull. Am. Phys. Soc. **8**, 62 (1963)
1.59   R.G. Pontin, E.F. Slade, D.J.E. Ingram: J. Phys. C**2**, 1146 (1969);
       O. Lewis, G. Wessel: Phys. Rev. B**13**, 2742 (1976)
1.60   U.T. Höchli: J. Phys. C**9**, L495 (1976)
1.61   T. Sakudo: J. Phys. Soc. Jpn. **18**, 1626 (1963)
1.62   T. Sakudo, H. Unoki: J. Phys. Soc. Jpn. **19**, 2109 (1964)

1.63   E. Siegel, K.A. Müller: Phys. Rev. B **20**, 3587 (1979)
1.64   T. Takeda: J. Phys. Soc. Jpn. **24**, 533 (1968)
1.65   R.M. Cotts, W.D. Knight: Phys. Rev. **96**, 1285 (1954)
1.66   See the discussion in Sect. 1.2, especially 1.2.4, of this chapter
1.67   See Chap. 2 on NMR of this volume
1.68   E. Siegel, W. Urban, K.A. Müller, E. Wiesendanger: Phys. Lett. **53A**, 415 (1975);
        E. Siegel: Ferroelectrics **13**, 385 (1976)
1.69   H. Ihrig: J. Phys. C **11**, 819 (1978)
1.70   K.A. Müller, W. Berlinger, J. Albers: Phys. Rev. B **32**, 5837 (1985)
1.71   J.P. Rivera, H. Bill, J. Weber, R. Lacroix, G. Hochstrasser, H. Schmid: Solid State Commun. **14**,
        21 (1974)
1.72   J.P. Rivera, H. Bill, R. Lacroix: Ferroelectrics **13**, 361 and 363 (1976)
1.73   K.A. Müller: Solid State Commun. **9**, 373 (1971)
1.74   H. Unoki, T. Sakudo: J. Phys. Soc. Jpn. **37**, 145 (1974)
1.75   M.O. Selme, P. Pecheur: J. Phys. C **18**, 551 (1985);
        F. Michel-Calendini, K.A. Müller: Solid State Commun. **40**, 255 (1981)
1.76   K.A. Müller, W. Berlinger, K.W. Blazey, J. Albers: Solid State Commun. **61**, 21 (1987)
1.77   I.B. Bersuker: Phys. Lett. **20**, 589 (1966)
1.78   K.A. Müller: Phys. Rev. Lett. **2**, 341 (1959)
1.79   R. Comes, M. Lambert, A. Guinier: Solid State Commun. **6**, 715 (1968)
1.80   K. Itoh, L.Z. Zeng, E. Nakamura, N. Mishima: Ferroelectrics **63**, 29 (1985)
1.81   K.A. Müller: Phys. Rev. B **13**, 3209 (1976); Ferroelectrics **13**, 381 (1976)
1.82   D. Rytz, L.A. Boatner, A. Châtelain, U.T. Höchli, K.A. Müller: Helv. Phys. Acta **51**, 430 (1978)
1.83   S. Triebwasser: Phys. Rev. **114**, 63 (1959)
1.84   See the Introduction in: *Topics in Current Physics*, Vol. 23, ed. by K.A. Müller, H. Thomas
        (Springer, Berlin, Heidelberg 1981), and the two texts by A.D. Bruce and R.A. Cowley with the
        title *Structural Phase Transitions* (Taylor and Francis, London 1981)
1.85   K.A. Müller, W. Berlinger: Phys. Rev. B **34**, 6130 (1986)
1.86   W.M. Walsh, J. Jeener, N. Bloembergen: Phys. Rev. **139A**, 1338 (1965)
1.87   R. Revai: Sov. Phys. Solid State **10**, 2984 (1969)
1.88   D. Rytz, U.T. Höchli, K.A. Müller, W. Berlinger, L.A. Boatner: J. Phys. C **15**, 3371 (1982)
1.89   R. Migoni, H. Bilz, D. Bäuerle: Phys. Rev. Lett. **37**, 1155 (1976)
1.90   M.D. Fontana, G. Métrat, J.L. Servoin, F. Gervais: J. Phys. C **16**, 483 (1984);
        M.D. Fontana, G.E. Kugel: Jpn. J. Appl. Phys. **24**, Suppl. 24–2, 223 (1985)
1.91   L. Castet-Mejean, F.M. Michel-Calendini: Ferroelectrics **37**, 503 (1981)
1.92   K.H. Weyrich, R. Siems: Z. Phys. B **61**, 63 (1986)
1.93   K.A. Müller: Helv. Phys. Acta **59**, 874 (1986);
        K.A. Müller: In *Nonlinearity in Condensed Matter*, ed. by A.R. Bishop, D.K. Campbell, P. Kumar,
        S.E. Trullinger, Springer Series in Solid-State Sciences, Vol. 69 (Springer, Berlin, Heidelberg
        1987) p. 235
1.94   H. Gränicher, K.A. Müller: Mater. Res. Bull. **6**, 977 (1971)
1.95   K.A. Müller, E. Brun, B. Derighetti, J.E. Drumheller, F. Waldner: Phys. Lett. **9**, 223 (1964)
1.96   H. Unoki, T. Sakudo: J. Phys. Soc. Jpn. **23**, 546 (1967)
1.97   K.A. Müller, W. Berlinger, F. Waldner: Phys. Rev. Lett. **21**, 814 (1968)
1.98   H. Thomas, K.A. Müller: Phys. Rev. Lett. **21**, 1256 (1968)
1.99   K.A. Müller, W. Berlinger: Phys. Rev. Lett. **26**, 13 (1971)
1.100   A. Aharony, A.D. Bruce: Phys. Rev. Lett. **42**, 462 (1979)
1.101   J.Y. Buzaré, J.C. Fayet, W. Berlinger, K.A. Müller: Phys. Rev. Lett. **42**, 465 (1979)
1.102   H. Gränicher: Helv. Phys. Acta **29**, 210 (1956)
1.103   W.I. Dobrov, R.F. Vieth, M.E. Browne: Phys. Rev. **115**, 79 (1959)
1.104   M.J. Buerger: In *Phase Transformations in Solids* (Wiley, New York 1951) p. 183; J. Chem. Phys.
        15, 1 (1947)
1.105   E. Pytte, J. Feder: Phys. Rev. **187**, 1077 (1969);
        J. Feder, E. Pytte: Phys. Rev. B **1**, 4803 (1970)
1.106   G. Shirane, Y. Yamada: Phys. Rev. **177**, 858 (1969)
1.107   P.A. Fleury, J.F. Scott, J.M. Worlock: Phys. Rev. Lett. **21**, 16 (1968)
1.108   G. Rupprecht, W.H. Winter: Phys. Rev. **155**, 1019 (1967)
1.109   A. Okazaki, M. Kawaminami: Mater. Res. Bull. **8**, 545 (1973)
1.110   H. Gränicher, K.A. Müller: Nuovo cimento Suppl. **6**, 1216 (1957)
1.111   W. Low, A. Zusman: Phys. Rev. **130**, 144 (1963)
1.112   S. Geller, V.B. Bala: Acta Cryst. **9**, 1019 (1956)

1.113   C. de Rango, S. Tsoucaris, C. Zelwer: Compt. Rend. **259**, 1537 (1964); Acta Cryst. **20**, 590 (1966)
1.114   V. Plakhty, W. Cochran: Phys. Stat. Solidi **29**, K81 (1968)
1.115   S. Geller, P.M. Raccah: Phys. Rev. B **2**, 1167 (1970)
1.116   F. Borsa, M.L. Crippa, B. Derighetti: Phys. Lett. **34A**, 5 (1971)
1.117   W. Cochran, A. Zia: Phys. Stat. Solidi **25**, 273 (1968)
1.118   J.D. Axe, G. Shirane, K.A. Müller: Phys. Rev. **183**, 820 (1969)
1.119   J.F. Scott: Phys. Rev. **183**, 823 (1969)
1.120   J.C. Slonczewski, H. Thomas: Phys. Rev. B **1**, 3599 (1970)
1.121   A.D. Bruce, A. Aharony: Phys. Rev. B **10**, 2078 (1974)
1.122   J.C. Slonczewski: Phys. Rev. B **2**, 4646 (1970)
1.123   L. Landau, E. Lifshitz: *Physique Statistique* (Editions MIR, Moscow 1967)
1.124   H.E. Stanley: *Introduction to Phase Transitions and Critical Phenomena* (Oxford Univ. Press, Oxford 1971)
1.125   S.K. Ma: *Modern Theory of Critical Phenomena* (W.A. Benjamin, London 1976)
1.126   V.L. Ginzburg: Fiz. Tverd. Tela **2**, 2031 (1960)
1.127   L.P. Kadanoff, W. Götze, D. Hamblen, R. Hecht, E.A.S. Lewis, V.V. Palciauskas, M. Rayl, J. Swift, D. Aspnes, J. Kane: Rev. Mod. Phys. **39**, 395 (1967)
1.128   M.E. Fisher: Rev. Mod. Phys. **46**, 597 (1974)
1.129   D.T. Taylor, W.D. Seward: Bull. Am. Phys. Soc. Ser. II **15**, 1624 (1970)
1.130   A.D. Bruce, A. Aharony: Phys. Rev. B **11**, 478 (1975)
1.131   K.A. Müller, W. Berlinger, M. Capizzi, H. Gränicher: Solid State Commun. **8**, 549 (1970)
1.132   I. Hatta, Y. Shiroishi, K.A. Müller, W. Berlinger: Phys. Rev. B **16**, 1138 (1977)
1.133   R. Bunde: Private communication
1.134   K.A. Müller, W. Berlinger: Phys. Rev. Lett. **35**, 1547 (1975)
1.135   J. Swift, M.K. Grover: Phys. Rev. A **9**, 2579 (1974)
1.136   F.J. Wegner: Phys. Rev. B **5**, 4529 (1972)
1.137   D.M. Saul, M. Wortis, D. Jasnow: Phys. Rev. B **11**, 2571 (1975)
1.138   M. D'Iorio, W. Berlinger, K.A. Müller: Phase Transitions **4**, 31 (1983)
1.139   K.A. Müller, W. Berlinger: unpublished (1975)
1.140   R.A. Cowley, A.D. Bruce: J. Phys. C **6**, L191 (1973)
1.141   G. Angelini, G. Bonera, A. Rigamonti: In *Magnetic Resonance and Related Phenomena*, ed. by V. Hovi (North-Holland, Amsterdam 1973) p. 346
1.142   K.A. Müller: *Nonlinear Phenomena at Phase Transitions and Instabilities*, Proc. NATO Advanced Study Institute, Geilo, Norway 1981, ed. by T. Riste (Plenum, New York 1981) pp. 1–34
1.143   K.A. Müller: In *Statics and Dynamics of Nonlinear Systems*, ed. by G. Benedek, H. Bilz, R. Zeyher, Springer Series in Solid-State Sciences, Vol. 47 (Springer, Berlin, Heidelberg 1983) pp. 68–79
1.144   P.A. Fleury, K. Lyons: In *Structural Phase Transitions I*, ed. by K.A. Müller, H. Thomas, Topics in Current Physics, Vol. 23 (Springer, Berlin, Heidelberg 1981) p. 32
1.145   M. D'Iorio, W. Berlinger, J.G. Bednorz, K.A. Müller: J. Phys. C **17**, 2293 (1984)
1.146a R.T. Harley, W. Hayes, A.M. Perry, S.R.P. Smith: J. Phys. C **8**, L123 (1975)
1.146b R.T. Harley: J. Phys. C **10**, L205 (1977) and references therein
1.147   R.J. Birgeneau, J.K. Kjems, G. Shirane, L.G. Van Uitert: Phys. Rev. B **10**, 2512 (1974)
1.148   P.A. Fleury, P.D. Lazay, L.G. Van Uitert: Phys. Rev. Lett. **33**, 492 (1974)
1.149   M.D. Sturge, E. Cohen, L.G. Van Uitert, R.P. van Stapele: Phys. Rev. B **11**, 4768 (1975)
1.150   M.E. Fisher, D.R. Nelson: Phys. Rev. Lett. **32**, 1350 (1974)
1.151   D.R. Nelson, J.M. Kosterlitz, M.E. Fisher: Phys. Rev. Lett. **33**, 813 (1974)
1.152   L.T. Todd, Jr.: MIT Crystal Physics Laboratory Technical Report 54 (1970)
1.153   V.H. Schmidt, A.B. Western, A.G. Baker: Phys. Rev. Lett. **37**, 839 (1976)
1.154   K. Fossheim, B. Berre: Phys. Rev. B **5**, 3292 (1972)
1.155   H. Rehwald: Solid State Commun. **21**, 667 (1977)
1.156   S. Stokka, K. Fossheim: Phys. Rev. B **25**, 4896 (1982)
1.157   K.A. Müller, W. Berlinger, J.E. Drumheller, J.G. Bednorz: In *Multicritical Phenomena*, ed. by R. Pynn, A. Skjeltorp, Proc. NATO Advanced Study Institute on Multicritical Phenomena, Geilo, Norway, 1983 (Plenum, New York 1984) p. 143
1.158   K.A. Müller, W. Berlinger, J.C. Slonczewski: Phys. Rev. Lett. **25**, 734 (1970)
1.159   A. Aharony, K.A. Müller, W. Berlinger: Phys. Rev. Lett. **38**, 33 (1977)
1.160   J. Rudnick: J. Phys. A **8**, 1125 (1975)
1.161   K.A. Müller, W. Berlinger: unpublished (1983);
        K.A. Müller: Ferroelectrics **53**, 101 (1984)
1.162   D. Blankschtein, A. Aharony: J. Phys. C **14**, 1919 (1981)
1.163   J.Y. Buzaré, J.C. Fayet: Solid State Commun. **21**, 1097 (1977)

1.164 P. Studzinski, J.M. Spaeth: J. Phys. C 19, 6441 (1986)

1.165a J.Y. Buzaré, M. Fayet-Bonnel, J.C. Fayet: J. Phys. C 13, 857 (1980)

1.165b J.Y. Buzaré: "Utilisations des paires $\{Fe^{3+}\text{-}O^{2-}\}$ et $\{Gd^{3+}\text{-}O^{2-}\}$ pour l'étude par R.P.E. de la transition structural $O_h^1 - D_{4h}^{18}$ des fluoperovskites..." Thesis. University of Le Mans (October 1978)

1.166 M. Arakawa: J. Phys. Soc. Jpn. 47, 523 (1979)

1.167 M. Fayet-Bonnel, J.Y. Buzaré, J.C. Fayet: Solid State Commun. 38, 37 (1981)

1.168 M. Fayet-Bonnel, A. Kaziba, J.Y. Buzaré, J.C. Fayet: Ferroelectrics 54, 281 (1984)

1.169 Y. Vaills, J.Y. Buzaré: J. Phys. C 20, 2149 (1987)

1.170 A. Bulou, C. Ridou, M. Rousseau, J. Nouet, A.W. Hewatt: J. Physique 41, 87 (1980)

1.171 P. Simon, J.J. Rousseau, J.Y. Buzaré: J. Phys. C 15, 5741 (1982)

1.172 C. Ridou, M. Rousseau, A. Freund: Solid State Commun. 35, 723 (1980)

1.173 E. Domany, D. Mukamel, M.E. Fisher: Phys. Rev. B 15, 5432 (1977)

1.174 K.A. Müller, W. Berlinger, J.Y. Buzaré, J.C. Fayet: Phys. Rev. B 21, 1763 (1980) and references therein

1.175 M. Kerzsberg, D. Mukamel: Phys. Rev. Lett. 43, 293 (1979); Phys. Rev. B 23, 3943 and 3953 (1981)

1.176 K. Fossheim: In *Multicritical Phenomena*, ed. by R. Pynn, A. Skjeltorp, Proc. NATO Advanded Study Institute on Multicritical Phenomena, Geilo, Norway, 1983 (Plenum, New York 1984), p.129

1.177 K.A. Müller, W. Berlinger: Z. Phys. B 46, 81 (1982) and references therein

1.178 R. Currat, K.A. Müller, W. Berlinger, F. Denoyer: Phys. Rev. B 17, 2973 (1978)

1.179 J.Y. Buzaré, W. Berlinger, K.A. Müller: J. Physique Lett. 46, L201 (1985) and references therein

1.180 A.D. Bruce, K.A. Müller, W. Berlinger: Phys. Rev. Lett. 42, 185 (1979) and references therein

1.181 M. Fujimoto, S. Jerzak, W. Windsch: Phys. Rev. B 34, 1668 (1986)

1.182 A. Leblé, J.J. Rousseau, J.C. Fayet, J. Pannetier, J.L. Fourquet, R. de Pape: Phys. Stat. Solidi 69, 249 (1982)

1.183a Y. Dagorn: "Etude par R.P.E. de la transition de phase ordre désordre dans NH_4AlF_4," Thesis. University of Caen (March 1984)

1.183b Y. Dagorn, A. Leblé, J.J. Rousseau, J.C. Fayet: J. Phys. C 18, 383 (1985)

1.184a A. Jouanneaux: "Etude par R.P.E. et diffraction de neutrons des composés mixtes $Rb_x NH_{4(1-x)}$," Thesis. University of Nantes (Dec. 1987)

1.184b A. Jouanneaux, A. Leblé, J.C. Fayet, J.L. Fourquet: Europhys. Lett. 3, 61 (1987)

1.185 A. Bulou, A. Leblé, A.W. Hewat, J.L. Fourquet: Mater. Res. Bull. 17, 391 (1982)

1.186 A. Bulou: Private communication

1.187 W. Selke, M. Barreto, J. Yeomans: J. Phys. C 18, L393 (1985)

1.188 A. Aharony: J. Magn. Mater 7, 198 (1978)

1.189 K.A. Müller: Ferroelectrics 72, 273 (1987)

1.190 R. Blinc: Phys. Reports 79, 331 (1981)

1.191 R. Blinc, J. Seliger, S. Zumer: J. Phys. C 18, 2313 (1985)

1.192 J. Emery, S. Hubert, J.C. Fayet: J. Physique 46, 2099 (1985)

1.193 V. Heine, J.D.C. McConnell: J. Phys. C 17, 1199 (1984)

1.194 P. Bak: Rep. Prog. Phys. 45, 587 (1982)

1.195 Y. Yamada, N. Hamaya: J. Phys. Soc. Jpn. 52, 3466 (1983)

1.196 H. Cailleau: In *Incommensurate Phases in Dielectrics – 2. Materials*, ed. by R. Blinc, A.P. Levanyuk, Modern Problems in Condensed Matters Science, Vol. 14.2, Series ed. by V.M. Agranovich, AA. Maradudin (North-Holland, Amsterdam 1986), p.71 and references therein

1.197 J. Dumas, R. Buder, J. Marcus, C. Schlenker, A. Janossy: Physica $143B$, 183 (1986)

1.198 J. Voit: Second European Workshop on CDW, Aussois (1987), Abstracts p.97

1.199 L. Bernard, R. Currat, P. Delamoye, C.M.E. Zeyen, S. Hubert, R. de Kouchkovsky: J. Phys. C 16, 433 (1983) and references therein

1.200 R. Currat, L. Bernard, P. Delamoye: In *Incommensurate Phases in Dielectrics – 2. Materials*, ed. by R. Blinc, A.P. Levanyuk, Modern Problems in Condensed Matters Science, Vol. 14.2, Series ed. by V.M. Agranovich, A.A. Maradudin (North-Holland, Amsterdam 1986), p.161 and references therein

1.201 N. Ait Yakoub, J. Emery, J.C. Fayet, S. Hubert: Solid State Commun. 60, 133 (1986)

1.202a N. Ait Yakoub: "Etude par R.P.E. de la transition de phase normal $\rightarrow$ incommensurable dans $ThCl_4$," Thesis. University of Le Mans (June 1987)

1.202b N. Ait Yakoub, J. Emery, J.C. Fayet, S. Hubert: J. Phys. C 21, 3001 (1988)

1.203 R. Blinc, D.C. Ailion, P. Prelovsek, V. Rutar: Phys. Rev. Lett. 50, 67 (1983)

1.204 G.F. Reiter, W. Berlinger, K.A. Müller, P. Heller: Phys. Rev. B 21, 1 (1980)

1.205 J.L. Ribeiro, J.C. Fayet, J. Emery, M. Pezeril, J. Albers, A. Klöpperpieper, A. Almeida, M.P. Chaves: J. Phys. (France) **49**, 813 (1988)
1.206 W. Brill, W. Schildkamp, J. Spilker: Z. Kristallogr. **172**, 281 (1985)
1.207 A. Klöpperpieper, H.J. Rother, J. Albers, H.E. Muser: Jpn. J. Appl. Phys. **24**, Suppl. 24–2, 829 (1985)
1.208 A.S. Chaves, R. Gazzinelli, R. Blinc: Solid State Commun. **37**, 123 (1981)
1.209 M. Pezeril, J. Emery, J.C. Fayet: J. Physique Lettres **41**, L499 (1980)
1.210 M. Pezeril, A. Kassiba, J.C. Fayet: unpublished work
1.211 A. Kaziba, J.C. Fayet: J. Physique **47**, 239 (1986)
1.212 A. Kassiba, M. Pezeril, P. Molinie, J.C. Fayet: Phys. Rev. B **40**, 54 (1989)
1.213 M. Pezeril, J.C. Fayet: J. Physique Lett. **43**, L267 (1982)
1.214 D. Durand, F. Denoyer: Phase Transitions **11**, 241 (1988)
1.215 A. Kaziba, M. Pezeril, J. Emery, J.C. Fayet: J. Physique Lett. **46**, L387 (1985)
1.216 J.J.L. Horiksc, A.F.M. Arts, H.W. de Wijn: Phys. Rev. B **37**, 7209 (1988)
1.217 M. Fujimoto, S. Jerzak: Philos. Mag. B **53**, 521 (1986)
1.218 A.S. Cullick, R.E. Gerkin: Chem. Phys. **23**, 217 (1977) and references therein
1.219 M. Fukui, S. Sumi, I. Hatta, R. Abe: Jpn. J. Appl. Phys. **19**, L559 (1980)
1.220 J.C. Toledano, P. Toledano: In *The Landau Theory of Phase Transitions: Application to Structural, Incommensurate, Magnetic, and Liquid Crystal Systems*, World Scientific Lecture Notes in Physics, Vol. 3 (World Scientific, Singapore 1987)
1.221 N. Merlin: "Forme de raies de R.P.E. dans les phases modulées [2ρ]" Thesis. University of Le Mans (December 1988)
1.222 J.C. Toledano, J. Schneck, C. Lamborelle: In *Symmetry and Broken Symmetries in Condensed Matter Physics*, ed. by N. Boccara (Inset, Paris 1981)

2. Comparison of NMR and NQR Studies of Phase Transitions in Disordered and Ordered Crystals

F. Borsa and A. Rigamonti

With 47 Panels

Since the development of microscopic theories for phase transitions in crystals, NMR–NQR techniques have been widely employed to study many of the static and dynamical phenomena occurring around T_c. For about two decades attention has been focussed on real systems approximating as closely as possible model systems characterized by translational invariance and described by simple model Hamiltonians, e.g. ferroelectric transitions in the quasi-harmonic phonon approximation (soft modes), order-disorder transitions in dipolar crystals in the pseudo-spin formalism, magnetic systems described by generalized Heisenberg models, etc. Within this framework, impurities, defects, domain walls, heterogeneity, etc., were basically regarded as unwanted effects limiting the applicability of the theories.

The discovery of particular effects such as discommensurations, central peaks, non-linear dynamics, relevant localized modes, and pretransitional clusters has progressively shifted interest towards disordered systems. Thus the emphasis is now on collective phenomena in systems where disorder and/or non-linearity are the major ingredients: spin glasses, dipolar, quadrupolar and orientational glasses, amorphous systems, charge density waves, incommensurate-commensurate transitions, adsorbates, intercalates, mixed crystals, etc. In these cases, magnetic resonance techniques have provided unique microscopic information not easily obtained by other experimental approaches. The main aim of this chapter is to present and discuss the most relevant contributions of NMR and NQR to the investigation of these "disordered" crystals. In doing this we will make frequent reference to the case of "ordered" crystals, where the understanding of NMR–NQR effects is fairly complete, as witnessed by numerous books and review articles. A complete review of the work carried out in the field is not in the purpose of this chapter. Instead, we will focus attention on some representative results which, in our opinion, best illustrate the uniqueness of the NMR–NQR approach.

2.1 Ordering in "Ordered" and Disordered Crystals

In reviewing NMR–NQR studies, it is appropriate to give a comprehensive scheme to illustrate the quantities that facilitate a microscopic description of the tendency of a many-body system to attain a state of order with spontaneous symmetry breaking. This will be done both for systems having lattice translational symmetry (which will be usually referred to as "ordered") and in systems with disorder either in the topology or in the interactions ("disordered").

We shall first recap on the well-known concepts and quantities involved in the usual description of phase transitions and critical phenomena. We then focus attention on less conventional systems where ordering and disordering phenomena are still partly unsettled. To simplify the presentation of a somewhat complex topic, some rather sweeping statements and generalizations will be necessary in some instances.

2.1.1 Phase Transitions and Critical Effects in Conventional Ordered Systems

Reversible structural phase transitions (SPT) in crystals belong to the wider class of phase transitions occurring as cooperative effects in many-body systems. The most general treatment is the statistical thermodynamical one, based on macroscopic order parameters and response functions [2.1, 2].

The onset of the order parameter (i.e. when it first becomes nonzero) below the transition temperature T_c marks a spontaneous symmetry breaking while the enhancement and slowing down of its fluctuations around T_c is reflected in the critical behavior of the appropriate response function.

In SPTs the above description is not, in general, possible. One may apply the Landau criterion for the definition of the order parameter and the order of the transition. According to this criterion the SPT is first order if a discontinuous change in the electric polarization and/or the lattice parameters occurs on cooling through the transition temperature: the two different crystallographic phases are in equilibrium with each other, with the same free energy, at the transition temperature. The transition is second order if the onset of the electric polarization and/or of an atomic displacement occurs continuously at the critical temperature. Although the reversible phase transitions considered here are characterized by a close orientation relationship between the two phases and by the occurrence of small and reversible atomic displacements, most of them are not second order, for example, the phase transitions in ferroelectrics, where an abrupt onset of the electric polarization is often observed. When the discontinuity in the order parameter is not too large, one treats the transition as quasi-continuous (or slightly first order) and the basic picture pertaining to critical phenomena in second-order transitions may be retained, provided that two characteristic temperatures are defined: the transition temperature T_c and the critical temperature T_0, the latter representing the stability limit for the thermodynamical functions. For a true second-order transition $T_c = T_0$.

For SPTs one can define as the order parameter an appropriate atomic displacement with respect to the lattice equilibrium positions in the high-temperature phase. For ferroelectric crystals it may be more convenient to refer directly to the spontaneous electric polarization (sublattice polarization for antiferroelectrics).

The interpretation of NMR and NQR effects at phase transitions in crystals is conveniently carried out in terms of local critical variables, $y(\boldsymbol{r}, t)$ (e.g. the orientation of a permanent dipole or the atomic displacement from an equilibrium position) and their average values, whose cooperative time behavior drives the transition. The relationships of the static critical response and of the dynamical structure factor with the correlation function for $y(\boldsymbol{r}, t)$ are summarized in Panel 2.1. The divergences in the thermodynamical quantities are found in correspondence to the collective variable

Generalities on Phase Transitions and Critical Phenomena

- Correlation function for the local critical variable and relationships with the response function.
- The pair correlation function for $y(r, t)$ is $G(r, t) = \langle y(r, t) y(0, 0) \rangle$ (equilibrium ensemble average) or $\mathrm{Tr}\{\varrho_0 y_i^* y_i'(t)\}$ (ϱ_0 equilibrium density matrix) when y is a quantum operator.
- The dynamical structure factor

$$S(q,\omega) = \int e^{i(\omega t - q \cdot r)} G(r, t) dr\, dt$$

describes the spectrum of the collective fluctuations.
- The spectrum of the fluctuations is related to the response function by the fluctuation-dissipation theorem:

$$S(q,\omega) = \frac{2\,\hbar \chi''(q,\omega)}{1 - \exp(-\hbar\omega/k_B T)} \simeq \frac{2 k_B T}{\omega} \chi''(q,\omega)$$

where $\chi''(q,\omega)$ is the dissipative part of the collective generalized susceptibility.
- The static structure factor

$$S(q) = \frac{1}{2\pi} \int S(q,\omega) d\omega = k_B T \chi(q, 0) = \langle |y_q|^2 \rangle$$

gives the mean-square amplitude of the collective fluctuations.
- The macroscopic order parameter (spontaneous polarization or cooperative lattice distortion) is related to the time average of $y(r_l, t)$ by:

$$\eta \text{ or } P \propto \sum_l e^{i q_c \cdot r_l} \langle y_l \rangle$$

where q_c represents the periodicity of the ordering in the local critical variable.
- Order parameter and static response function around the transition temperature.

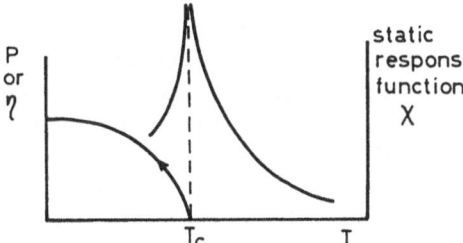

Fig. 2.1. For a second-order phase transition the order parameter goes to zero as $P \propto (T_c - T)^\beta$. Correspondingly the static susceptibility $\chi(q,0)$ for the critical wave vector q_c diverges as $\chi(q_c, 0) \propto |T - T_c|^{-\gamma}$

at the critical wavevector q_c, representing the periodicity of the staggered ordering in the local critical variables below T_c. For fluids as well as ferromagnets $q_c = 0$. In SPTs in ordered crystals, in general, q_c is restricted to high symmetry points in the Brillouin zone (for example, $q_c = 0$ for ferroelectrics, and q_c is at the zone boundary for antiferrodistortive transitions).

On approaching the critical temperature in a second-order transition, the static response function $\chi(q_c, 0)$ for a given critical wave vector q_c diverges, thereby reflecting the enhancement of the fluctuations of the collective variables (Fig. 2.1). For ferroelectric systems the appropriate critical response function is the dielectric susceptibility. In a pure SPT, one defines a susceptibility that represents the response of the system to the application of a generalized force which is the thermodynamic conjugate of the order parameter and may not have an intuitive physical realization.

The temperature dependence of the order parameter can be obtained in the framework of thermodynamical Landau-type theories or microscopic mean-field treatments. Since the time average of the critical variable y_l can be directly related to the order parameter, we set $\langle y_l \rangle \propto \varepsilon^{\beta}$ where $\varepsilon = (T_0 - T)/T_0$ (T_0 coinciding with T_c for a second-order phase transition), and β is the critical exponent for the temperature dependence of the order parameter. In the Landau-type or mean-field theories $\beta = 0.5$. Close to the transition temperature a non-classical critical region occurs, in which the mean-field approximation (MFA) fails. This region covers a temperature interval that depends on the range of the interaction driving the phase transition and on the lattice dimensionality. The value of β in the non-classical critical region is expected to be largely independent of the nature of the microscopic forces, depending instead on the dimensionalities and on the symmetries of the system and of the critical variable.

The general features of the time-dependent phenomena can be illustrated by recalling some properties of $S(q, \omega)$. A rigorous classification for SPTs is possible on the basis of the single-particle local potential $V(y)$, in terms of a one-dimensional spatial coordinate [2.3, 4]. In Panel 2.2 we refer, for illustrative purposes, to a simple model of a local anharmonic potential which leads, in the two limiting situations, to the classes of *displacive* and *order–disorder* SPTs.

A valuable criterion for the classification of the type of SPT can also refer to the character of the ordering: the first class is characterized by a continuous cooperative displacement at T_c along y, (at least in the mean-field approximation) while the second yields an ordering in the occupation of the two deep minima.

One may separate the correlation function for the collective variable into a static and a time-dependent part. Thus one may illustrate separately the general properties of the ω-integrated $S(q, \omega)$, namely the static structure factor, and the ω-distribution of the normalized quantity $J_q(\omega) = S(q, \omega)/S(q)$ (Panel 2.3).

The microscopic correspondence of the thermodynamical enhancement of the fluctuations is given by the divergence, on approaching T_c, of $S(q_c)$. The slowing down of the fluctuations is reflected, instead, in the progressive increase of the low-frequency components in $J_{q_c}(\omega)$.

Close to the transition, and for q close to q_c, the general properties of $J_q(\omega)$ may be predicted on the basis of dynamical scaling [2.5], whereby for $T > T_c$ the

Local Potentials and Dynamical Structure Factor

- Simple model of local potentials leading to displacive and order-disorder SPTs respectively (Fig. 2.2a) and corresponding dynamical structure factors (Fig. 2.2b).

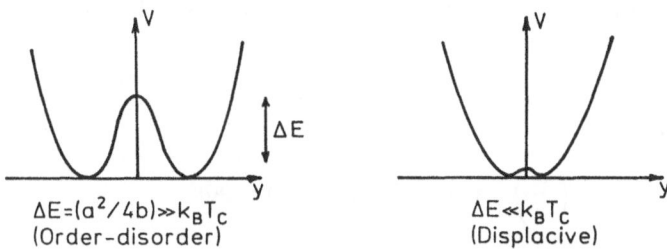

$\Delta E = (a^2/4b) \gg k_B T_c$
(Order-disorder)

$\Delta E \ll k_B T_c$
(Displacive)

Fig. 2.2a. Local potential $V(y) = ay^2 + by^4 (a < 0, b > 0)$

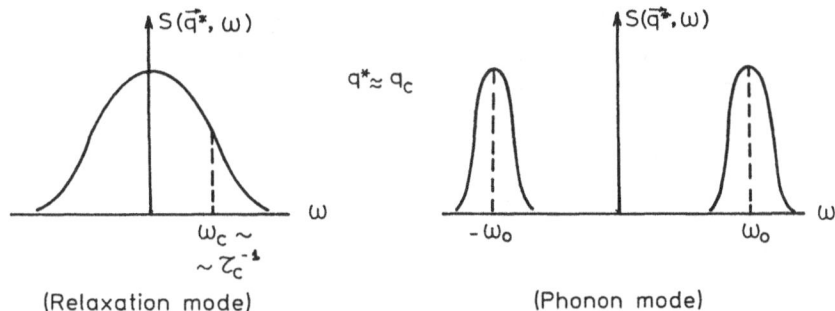

$q^* \approx q_c$

(Relaxation mode)

(Phonon mode)

Fig. 2.2b. Dynamical structure factors $S(q, \omega)$ for order-disorder transitions (typically of relaxational character) and for displacive transition (of resonant character, for T not too close to T_c)

General Properties of the Static and Dynamical Structure Factors Around a Transition

- On approaching the critical temperature, the enhancement and slowing down of the fluctuations are reflected in $S(q, \omega)$. The general aspects of the critical behavior can be described in terms of static and dynamical scaling:
- For the critical wave vector q_c, the area of $S(q_c, \omega)$ diverges with the critical exponent γ:

$$S(q_c) = k_B T \chi_0 \varepsilon^{-\gamma} \quad ; \qquad \varepsilon \equiv \frac{(T - T_c)}{T_c} \quad .$$

- The q-dependence of $S(q)$, for $q \approx q_c$, can be expressed in terms of a characteristic length scale $\xi \equiv \kappa^{-1}$

$$S(q) = S(q_c)\kappa^2 / (q^2 + \kappa^2 + \text{corr. terms}) \quad ; \qquad \kappa = \kappa_0 \epsilon^\nu$$

which is a particular form (corrected MFA) of the static scaling.
- The general behavior of the normalized spectrum of the fluctuations $J_q \equiv S(q,\omega)/S(q)$ consists in a progressive shift of its components towards the low-ω range; for $T \gtrsim T_c$ and q close to q_c, one can write

$$J_q(\omega) = \frac{2\pi}{\Gamma_q} f\left(\frac{\omega}{\Gamma_q}\right)$$

$$\Gamma_q(T) \simeq Aq^z g'\left(\frac{q}{\kappa}\right) = \kappa^z g\left(\frac{q}{\kappa}\right) \quad \text{with} \quad \Gamma_q(T \simeq T_c) = Aq^z \quad .$$

- In the dynamical MFA, $z = 2$ and

$$g\left(\frac{q}{\kappa}\right) = \Gamma_0 \kappa_0^{-2}\left[1 + \left(\frac{q}{\kappa}\right)^2\right] \quad .$$

- Illustration of the effect of the slowing down around T_c for order-disorder transition and for displacive-type transition.

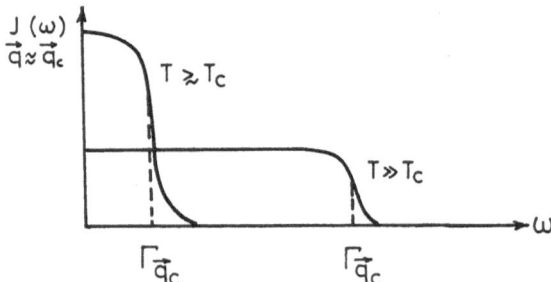

$$\Gamma_{\vec{q}_c}^2 \propto (T - T_c)$$

Fig. 2.3a. Order-disorder transition

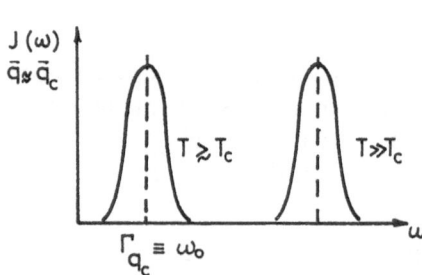

Fig. 2.3b. Displacive transition

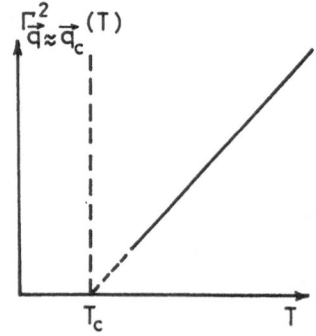

$$\Gamma_{\vec{q} \approx \vec{q}_c}^2(T)$$

spectral function depends in an essential way on the ratio q/κ, κ being the inverse correlation length.

2.1.2 Ordering Phenomena and Dynamics in Disordered Systems

In the previous section it was shown how structural phase transitions in ordered systems are characterized by the occurrence of an order parameter and by a peculiar critical dynamics, involving a slowing-down of a collective mode of relaxational and/or resonant character.

This picture is altered in essential ways when disorder of a different type is present. While in ideal non-disordered systems there is, in general, a precise relationship between local critical variables and macroscopic quantities and response functions, this relationship is more hidden when disorder is present. In this case, techniques like NMR–NQR are particularly powerful since the "local" dynamics and the "local" ordering effects [2.6] become more relevant here. This explains the major contribution that NMR–NQR is making to the understanding of phenomena such as random field effects, nature and formation of glass-like phases, freezing of molecular orientations, metastability, etc.

An example of disorder is obtained when in an ideal crystal, which undergoes a phase transition, one produces a random atomic substitution, such as the substitution of the "ferroelectric" ion Nb by Ta in $KNbO_3$. In this case the transition temperature T_c becomes concentration dependent and the major interest is in static and dynamical effects around the critical concentration defining the percolation threshold. Another example of disorder is an assembly of electric dipoles randomly located on a regular crystal lattice, such as the off-center Li ions substituting for K in the paraelectric $KTaO_3$ crystal. One of the most appealing features of such a system is the similarity to magnetic spin glasses, with the advantage that several experimental approaches such as dielectric, ultrasonic, and optical techniques are applicable in addition to NMR–NQR. An analogous system is represented, for example, by the mixed $(KCN)_x(KBr)_{1-x}$ crystals, where the randomly located CN groups are mostly elastically interacting quadrupoles which can form an orientationally disordered phase. Finally, it is worth mentioning that isotopic disorder is particularly relevant for hydrogen bonded ferroelectric crystals such as the KDP family: substitution with deuterium drastically affects the tunneling integral, thus changing the quantum tunneling contribution to the reorientation of the local dipole in the double-well potential.

In NMR–NQR studies of disordered systems the relevant quantities are the average of the local variables and the long-time dynamics. These quantities are crucial in the understanding of "freezing" as a cooperative phenomenon [2.7]. The definition of an order parameter in terms of the average local quantities for systems lacking translational invariance is not trivial. The Edwards-Anderson order parameter [2.8, 9] is defined as $q_{EA} = (1/N)\sum_i \langle s_i \rangle^2_{T,J}$, where the local critical variable s_i is first thermally averaged and subsequently averaged over the distribution of the interactions. Thus $\sqrt{q_{EA}}$ can be viewed as the "frozen" local effective moment per lattice site. NMR yields, in principle, the distribution of $\langle s_i \rangle_T$ through the details of the spectrum (Panel 2.4).

Static and Dynamical Critical Effects in Disordered Systems

- The definition of order parameters in disordered systems can be given by
  - i) $\sqrt{q_{EA}} \equiv \langle|\langle s_i\rangle_{t_m}|\rangle \neq 0$, where $\langle\ \rangle$ is an average over the time-scale of the experiment and a further spatial averaging is then performed;
  - ii) $\langle\langle s_i\rangle \cdot \langle s_j\rangle\rangle \neq 0$ for $|R_i - R_j| \leq \lambda$, indicating preferential orientation within the spatial dimension λ of a correlated cluster, while

$$\lim_{(R_i - R_j)\to\infty} \langle\langle s_i\rangle \cdot \langle s_j\rangle\rangle = 0$$

  namely there is no long-range preferential orientation.

- Correspondingly, in terms of the auto-correlation function, one has the behavior of Fig. 2.4a: the simple freezing implies that $g(t)$ is flat over the time-scale of the observation, while the system is in a glassy state when g (flat) is greater than the statistical equilibrium average taken over all the configurations (non-ergodicity).

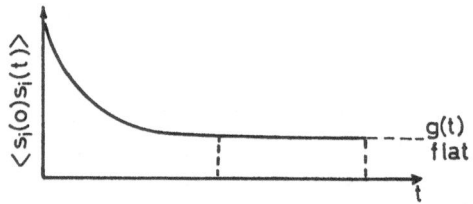

Fig. 2.4a. Behavior of the auto-correlation function $g(t)$

- The critical dynamics in disordered systems, namely the equivalent of the mode softening in ordered crystals, is often described by a distribution of local disordering modes of relaxational character, as depicted in Fig. 2.4b; when tunneling is present, in general one can have relaxational modes for phonon-assisted tunneling (asymmetric double-well potential) or resonant modes (symmetric double-well potential) (see Fig. 2.4c).

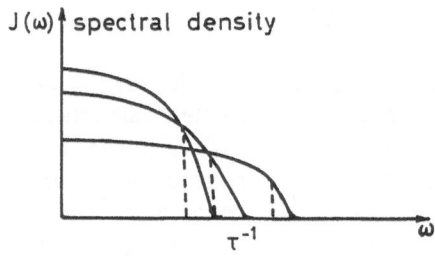

Fig. 2.4b. Distribution of correlation frequencies τ^{-1}: the freezing process is driven by relaxational modes, with a distribution of effective correlation times; the average is described by the empirical Fulcher law $\langle\tau\rangle \propto \exp[c/(T - T_0)]$

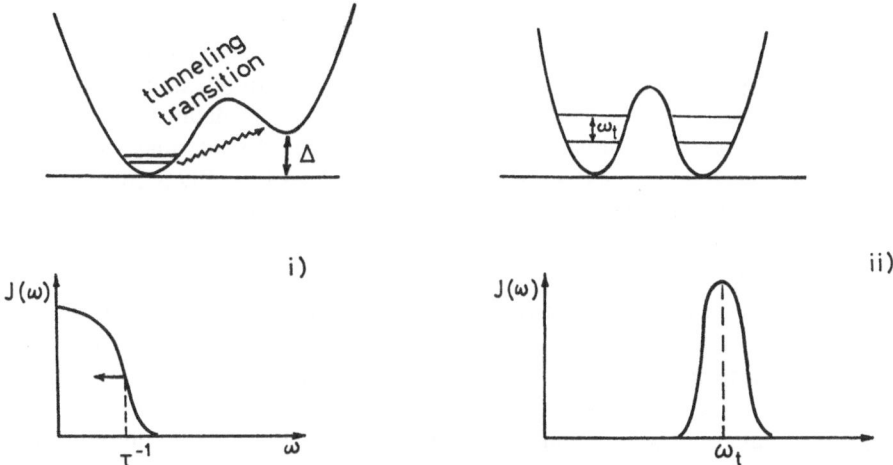

Fig. 2.4c. (*i*) Phonon-assisted tunneling yields relaxational modes with an effective correlation frequency given by the transition rate (disorder modes); (*ii*) Resonant tunneling yields resonant modes associated with tunneling splitting

The cooperative slowing-down in disordered systems may sometimes not lead to a thermodynamical phase transition. In this case an apparent order parameter might show up because, during the time-scale t_m of an experiment, the correlation function of the local variable keeps a practically constant value. In order to accommodate the possibility that the low temperature phase is only dynamically correlated over long times, one can define an order parameter in terms of the time-dependent correlation function [2.10] $q(t_x) = \langle\langle s_i(0)s_i(t_x)\rangle_T\rangle_J$. For $t_x = t_1$ equal to the shortest of the time scales, which become infinite in the thermodynamical limit, $q(t_1) \equiv q_{EA}$.

An apparent frozen phase may be observed over the observation time t_m in all cases. Again NMR experiments, using both spectra and relaxation rates, are very suitable in view of the possibility of moving t_m to much longer times than other types of experiments. Furthermore, although the average local quantities and, where present, the macroscopic order parameter are not simply connected, the analysis of the NMR data yields the spatial distribution of the local quantities averaged over a time-scale t_m. This distribution is the only experimentally accessible parameter for the description of the freezing process in the presence of many-body interactions and frustration.

We now mention briefly the main characteristics of the critical dynamics in disordered systems [2.7], i.e. the phenomena which are the equivalent of mode softening for SPTs in ordered crystals.

In some instances, crystals with random atomic or isotopic substitution and binary mixed crystals retain the general properties of the pure system. Here, in the linear approximation, one still has soft modes with eigenfrequencies ω_1 and ω_2 of the two coupled modes that are functions of the concentration. The working approach is often the dynamical random MFA approximation. Examples of this kind are the isotopically mixed ferroelectrics KDP-DKDP.

In the dilute limit, when the random MFA is expected to break down, one has to take into account the occurrence of local modes. These modes can also change their character going from resonant modes at high temperature to relaxational modes below a "critical temperature". A system of this type, treated in detail in Sect. 2.4, is $KTaO_3$ with random substitution of K by Na. Furthermore, the interplay of glass-type local modes and the soft modes can give rise to interesting effects like "local freezing", metastability, etc. In disordered systems where the random distribution of conjugate fields produces frustration, the soft-mode concept has to be abandoned. The single-particle dynamics is normally relaxational and the cooperativity causes departures from the Arrhenius-type temperature behavior and/or departures from the exponential recovery of the response functions (e.g. the electric polarization in the time-domain measurements). The departure from the Arrhenius behavior with a single correlation time can give rise to a distribution of correlation times whose width increases critically, and/or a temperature dependence described by the Fulcher law (Panel 2.4).

Finally one can mention the case in which tunneling of an atom or a group of atoms between two minima of a double well potential is a relevant part of the dynamics. The distribution of strains and of elastic or electric interactions may lead to a random asymmetry Δ in the levels of the two wells with a wide distribution of Δ's. At low temperature one can have peculiar dynamics for a limited number of units, due to transitions from one well to the other promoted by non-resonant phonon-assisted tunneling (disorder modes). These modes are believed to be responsible for the low-temperature linear behavior of the specific heat in glasses [2.11]. For equivalent minima, ($\Delta = 0$), the resonant tunneling with a characteristic frequency ω_t may become relevant. The random distribution of local barriers may lead to a wide distribution of tunneling frequencies and consequently to a low temperature spectral density of the motions described by a Lorentzian type function centered at ω_t, with a width critically T-dependent. This feature can be seen as a random statistical distribution of resonant tunneling δ-type transitions (Panel 2.4).

2.1.3 Non-linear Dynamics:
Anharmonicity, Damping, Solitons, and Tunneling

In this section we briefly describe some features of non-linearity that specifically affect the low-frequency part of the spectral density of motion. To a certain extent, non-linear effects are always present at phase transitions, due to the large fluctuations near the critical point (Sect. 2.1.1). In disordered systems non-linearity causes strong dynamical effects even in situations where no phase transitions occur. For displacive-type systems far from the transition, one can apply the usual description for the atomic motions in terms of harmonic lattice vibrations [2.4]. The normal modes can be easily singled out and labelled according to symmetry conditions. The lattice with which the resonant nuclei are in thermal equilibrium is a "quantum lattice" and exchange of energy between nuclei and phonons involves the absorption or emission of a low-frequency quantum $\hbar\omega_L$. In the linear approximation the process can occur with one-phonon direct exchange or a two-phonon Raman process. In the first case low-energy phonons are involved, whose density of states is negligibly small in three dimensions. Thus, as we will see in Sect. 2.2, in the absence of sizeable non-linearity, the most effective nuclear relaxation mechanism is the two-phonon indirect process. In quasi-harmonic theories, small non-linear effects are taken into account by properly renormalizing the elastic constants which become temperature dependent. Furthermore, the effects on the dynamics can be described in terms of a damping factor Γ which limits the life-times of harmonic phonons. An approximate way to include this effect in the spectral density is to replace the δ-like Fourier components of the motion with Lorentzians of width Γ.

In the presence of strong damping the low-frequency range of the spectral density becomes typical of overdamped, diffusive-like modes. Here the lattice thermal bath can be treated classically as for random motions, and the characteristic correlation frequencies are of the order of the damping factor. One should emphasize, at this point, that the non-linearity considered here for displacive phase transitions originates from the terms in the equations of motion which are neglected in the quasi-harmonic approximation (e.g. phonon-phonon interactions).

In ideal order-disorder systems the critical variable corresponds to large off-center displacements of atoms or groups of atoms which obviously cannot be treated by linear expansions [2.4]. The single-particle behavior is of a Debye-type with a correlation time τ_0, while basically two approaches can be used to describe the collective behavior. In the approach based on the Glauber master equations, the q-dependent correlation time for the diffusive collective mode is a function of τ_0 and of the Fourier transform of the interaction energy $I(q)$. The second approach is the pseudo-spin model, which also takes into account the possibility of quantum tunneling between the double minima of the local potential. The nature of the critical modes depends on the interplay of diffusive thermal fluctuations of Debye-type and resonant quantum fluctuations. Up to here the situation described is the one pictured in Panel 2.2 for order-disorder systems. Non-linear effects have to be viewed as leading to modifications to the above picture. In particular, one can have a damping of the tunneling modes due to the coupling with phonons, or effects related to higher-order terms usually neglected in the Glauber or pseudo-spin models, causing a variety of modifications in the low frequency parts of the spectral density (see also the discussion of the central peak in the next section). A model in which one tries to include the complete form of the local potentials (as depicted in Panel 2.2) without linearization is the one-dimensional model of elastically coupled particles in local double minimum potentials [2.12, 13]. By solving the equation of motion, including the inertia terms, in the classical continuum limit, one finds the ordinary phonon-like excitations and soliton-type solutions which correspond to a particle jumping over the barrier involving a few neighboring particles and propagating coherently along the chain. It is convenient to think of the phonon solutions as the linear excitations and the soliton-type kinks as the excitations incorporating the whole non-linearity.

As we will discuss further, the solitons can be viewed as domain-walls of the pretransitional clusters, which give rise to the central peak in displacive-type phase transitions. A situation where the non-linearity gives rise to particularly relevant soliton-type effects is that found in incommensurate systems, as depicted in the simplest way by the 1D incommensurate model with a sinusoidal potential well, leading to Sine-Gordon soliton solutions [2.14] (Panel 2.5).

Non-linear Effects in SPTs

- A simple model of non-linear dynamics applicable to SPTs consists in a one-dimensional array of particles in a double-well potential with nearest neighbor bilinear coupling (Fig. 2.5a).

  The Lagrangian function is

$$L = \frac{m}{2} \sum_n \dot{s}_n^2 - \sum_n \left[V_0 \left(1 - s_n^2\right)^2 + \frac{c}{2} \left(s_n - s_{n+1}\right)^2 \right]$$

  and in the classical continuum limit the equation of motion

$$m\ddot{s} - 4V_0 \left(1 - s^2\right)s - ca\frac{\partial^2 s}{\partial x^2} = 0$$

  leads to kink-like solutions for the displacement $s = s(x, t)$ (Fig. 2.5b).
- An approximate form for the low-frequency spectral density is shown in Fig. 2.5c; when $k_B T \ll V_0$, the correlation time $\tau = \xi/v$ (ξ correlation length or soliton width, v around the sound velocity) is approximately of the form $\tau \propto \exp(V_0/k_B T)$.

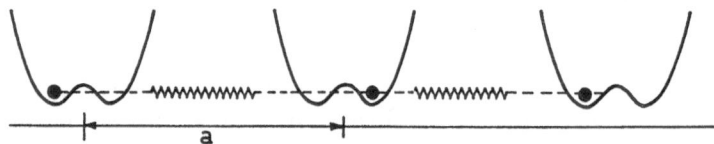

Fig. 2.5a. The model of 1D interacting particles

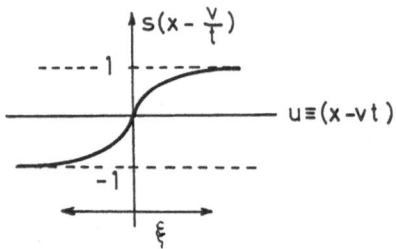

Fig. 2.5b. Soliton kink-like solution (ξ correlation length)

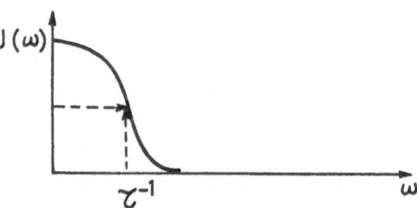

Fig. 2.5c. Low-frequency spectral density

We mention here the effect of quantum tunneling on the spectral density of order-disorder motions, even though, strictly speaking, it cannot be considered a non-linear phenomenon. We have seen that in a double-well potential and for $T \ll \Delta E/k_B$, hopping of an ion or reorientation of a dipole is the classical thermally activated process giving rise to a spectral density centered at zero frequency and width τ^{-1}. When the ions are light (typically hydrogen) or the two minima are very close, quantum tunneling can contribute to the dynamics. In the pseudospin model [2.4] the quantum tunneling is taken into account through the S_x spin-operator and introduces an intrinsic dynamical term in the Hamiltonian, yielding collective motion of resonant character, in the form of pseudospin-waves with quasi δ-like spectral densities centered at a q-dependent frequency. The coupling of the tunneling with the phonon bath introduces complicated effects that can be viewed as non-linear. In particular, damping can change the character of the excitation, leading to a low-frequency relaxational-type spectral density, as in the dynamical Ising model (Glauber model).

In systems lacking translational invariance, the description of the dynamics in terms of linear excitations is usually difficult. The simplest approach is to take averages over the configurations and to replace the random variable and/or interactions by average quantities in the Hamiltonians. The general consensus is that non-linear effects such as domain wall propagation, kinks, defects, and localized modes become more relevant, as we will see in discussing more specific cases.

In view of the importance to spectroscopic studies of a particular effect, namely, non-linear dynamics consisting in a central component in the spectral densities, superimposed on the contribution from the ordinary linear excitations, we attempt in the following a clarification of the problem.

2.1.4 Central Peak

One can envisage *extrinsic* and *intrinsic* mechanisms which cause the appearance of a central peak (CP).

A mechanism is called *extrinsic* if it is related to the presence of impurities, surface effects, local breaking of symmetries which should ideally be absent for a pure, infinitely large, translationally invariant single crystal. The *intrinsic* mechanisms are the ones due to non-linear terms in the Hamiltonians, which are also present in the ideal crystal. One can recognize at least three different intrinsic mechanisms leading to the occurrence of a central peak. The "classical" central peak is related to non-linear excitations generated in the vicinity of a second-order phase transition. Here large fluctuations occur, in principle both for displacive and order–disorder type phase transitions, and the excitations related to the development of short range order are often described in terms of solitons or domain-wall motions. CPs of this type have been observed in computer simulations. In the second type of intrinsic central peak we include the heterophase fluctuations or thermodynamical precursor effects which are also ideally present without impurities. These effects are expected at first-order phase transitions and might be very difficult to detect. They should cause the appearance of pretransitional clusters with long-range order. It may be mentioned that, in this second case, the thermal bath plays a crucial role because it must provide the energy for the thermal hopping in the metastable thermodynamical state. This situation should be contrasted with the first case where the thermal bath exchanges energy with the ordinary elementary excitations.

Finally it should be noted that the presence of impurities, defects, or strains can affect the dynamics of intrinsic-type pretransitional clusters (either of the short-range order related to non-linear dynamics, or of long-range order) (Panel 2.6).

Since in real systems impurities and/or defects are always present, the experimental observability of a purely intrinsic CP is always difficult. However, the difference between extrinsic and intrinsic CPs still remains, inasmuch as the former introduces an *additional* contribution to the total spectral density of the intrinsic excitations. The latter, even in the presence of pinning and interaction with defects, introduces a modification in the low-frequency spectral width without changing the total area.

Next we give some illustrative examples for the different types of CPs, referring to theoretical descriptions in which the phenomenon clearly results from a particular microscopic model. Once again we stress that we choose only some examples from a copious literature in the field.

An extrinsic mechanism for a CP has been described by *Halperin* and *Varma* [2.15]. In this model a relaxing defect cell is introduced, which should correspond to a lattice imperfection. A model Hamiltonian is set up where the local potential is assumed to be different in the normal and defective cell. In this way one obtains a soft-mode behavior for the normal cells and superimposed on it a local dynamics, on a much slower scale, corresponding to the thermal hopping between equivalent displaced positions of the defect in the cell. Another model accounting for a CP of extrinsic nature is that in which the defects cause a local phonon condensation and the resulting cluster couples linearly to the soft phonons [2.16].

Central peaks of intrinsic origin at second-order phase transitions are clearly evident in computer simulation experiments [2.17]. The related mechanisms have been called variously "solitons", "domain-wall motions", "pretransitional clusters", etc. Many theoretical treatments have proposed microscopic models leading to CP dynamics in addition to the soft modes. One type of approach relies on the classical solution of the non-linear field equations which describe the strongly anharmonic motions of the atoms. In the treatments pioneered by *Krumhansl* and *Schrieffer* [2.12], *Varma* [2.18] and *Aubry* [2.13], the CP is basically due to the onset of a double-well shaped local potential for interacting atoms (Panel 2.5). The collective motion of the atoms results in the superposition of phonon modes and soliton-like modes corresponding to the propagation of the boundary between the clusters of short-range order. Similar results are also obtained by treating the dynamics of a limited number of coupled anharmonic oscillators [2.13, 19]. In another group of theoretical approaches the CP is obtained within the traditional framework of renormalized phonon theory by introducing a frequency-dependent damping factor into the equation of motion for quasi-harmonic oscillators [2.20], or, equivalently, in the phonon self-energy [2.21]. It is worth stressing, at this point, that the difficulty in detecting a long-time scale dynamics is present only when the critical dynamics is related to the softening of lattice phonon modes [2.22]. In this case, in fact, the occurrence of a CP involves a cross-over from displacive to order-disorder critical motion. In order-disorder phase transitions, such as in $NaNO_2$, the critical slowing

Central Peaks

- The CP in displacive SPTs can be of *extrinsic* character (impurities and/or defects) or of *intrinsic* character when generated by non-linear effects (see Panel 2.5 for a simple 1D model).

    The intrinsic CP is characterized by a phonon softening, accompanied by a growing central component in $S(q, \omega)$ (Fig. 2.6a). For order-disorder SPTs the CP is superimposed to the central component in $S(q, \omega)$ related to the relaxational mode of the order-disorder variable (Fig. 2.6b).

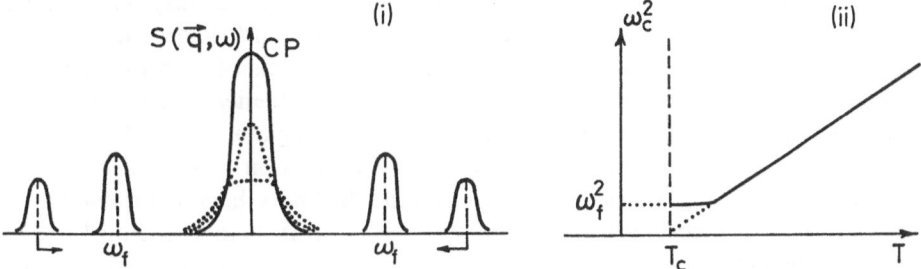

Fig. 2.6a. Dynamical structure factor $S(q, \omega)$ for $q \simeq q_c$, (*i*), and temperature behavior of the critical frequency (*ii*) for a system displaying an intrinsic CP. Note that the softening levels off at ω_f while the central component in $S(q, \omega)$ diverges critically

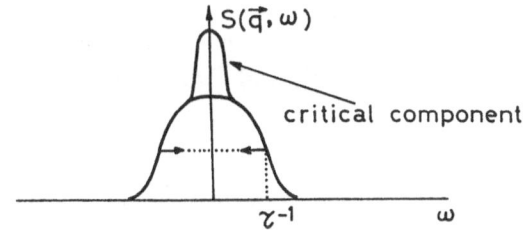

Fig. 2.6b. Typical structure of $S(q, \omega)$ for an order-disorder system showing the CP superimposed to the ordinary dynamical structure factor of relaxational character (Panel 2.2)

---

of the order-disorder variable leads to a CP in a natural way. Strictly speaking, one should consider the critical relaxational mode as the equivalent of the soft mode rather than of the CP, unless pinning, coupling to defects, or a particularly long-time tail in the correlation function, modifies the low-frequency part of the relaxational mode. The non-critical short time scale associated with quasi-harmonic phonon motion is irrelevant here except when the coupling induces local effects; these are particularly important in NMR–NQR approaches, as will be seen.

In magnetic systems the central component is always present in the paramagnetic phase. In 1D systems, for example, a critical central component is observed in a wide temperature range and corresponds to the build-up of short-range correlations.

A fundamentally different intrinsic mechanism for a CP is provided by the heterophase fluctuations in first-order phase transitions. By extending the liquid droplet model to crystals, *Cook* [2.23] has emphasized that a pretransitional cluster can form when, on cooling, the free energy of the high temperature phase approaches the free energy of the metastable low temperature phase. The width of the resulting CP is controlled by the time required for the formation and collapse of the clusters. Analogous to the gas-liquid transition, the occurrence of a CP due to heterophase fluctuations is strictly limited by the strain energy contribution associated with the formation of the long-range order clusters.

Regarding the experimental evidence, some controversies have arisen due to the difficulties in separating static and dynamical effects in measuring the width of the CP (often within the resolution limits in most scattering experiments), and because of the different regions of the crystal (surface or bulk) probed by different techniques. Light scattering experiments have been interpreted in terms of both extrinsic (static and dynamical) and intrinsic CPs [2.24]. X-ray critical scattering and γ-ray recoilless scattering, which probe mainly perturbed surface regions, should be considered as providing information mostly on extrinsic CPs. Neutron scattering detects bulk regions as well as regions around defects and the data have been interpreted both in terms of extrinsic mechanisms [2.16] and in terms of intrinsic mechanisms [2.25–27]. An unambiguous assignment of a CP to a precise microscopic mechanism depends crucially on the frequency width of the peak, a measurement which is usually made difficult by resolution limitations in neutron scattering studies [2.28, 29].

From the frequency dependence of the ultrasonic attenuation coefficient, the occurrence of an intrinsic CP has been claimed in $KMnF_3$, and the frequency width and its critical narrowing have been evaluated [2.30].

NMR and dielectric low-frequency dispersion measurements give, in principle, the most direct measure of the central components in $S(q, \omega)$ (Sects. 2.3 and 2.4).

2.1.5 Incommensurate Phases

As mentioned in Sect. 2.1.1, the spatial variation of the average of the local critical variable (a kind of "local order parameter") is described by a critical wave vector q_c corresponding to points of high symmetry in the Brillouin zone, and the new unit cell in the ordered phase is obtained as a simple multiple of the high temperature phase. In the framework of the thermodynamical description of phase transitions (based on the expansion of the free energy density in terms of powers of the order parameter and of its gradients) [2.1], a commensurate phase corresponds to the case where the free energy is minimized below T_c by an order parameter with critical wave vector q_c at a point of high symmetry in the Brillouin zone [2.31].

An incommensurate phase, on the other hand, occurs for an order parameter with a spatial periodicity which is not commensurate with that of the underlying crystal lattice. The translational invariance is thus broken at T_c and the structure cannot be described, in general, by an ordinary space group.

A description of the phenomenon can be obtained by postulating specific thermodynamic potentials of the Landau-Lifshitz type, although they must clearly be compatible with the physics of the system. From a microscopic point of view, the mechanism generating an incommensurate phase, generally difficult to describe, is often believed to be the competition between interactions of different ranges (for nearest neighbor interactions no incommensurate phase can occur). A semi-macroscopic description is the generalization of the soft-mode theory for ordinary SPTs, with a quasi-harmonic potential. The generalization is obtained through a renormalization-group approach based on Landau-Ginzburg-Wilson-type effective Hamiltonians [2.32]. Competition between interactions can also lead to a devil's staircase of configurations, as exemplified in NH_4AlF_4-$RbAlF_4$ mixed crystals [2.33].

In metallic systems, incommensurate phases can be generated, particularly for low-dimensional systems, by the onset of a modulation of the conduction electron charge density (charge density wave or CDW), associated with the particular topology of the Fermi surface. Incommensurate phases have been also observed in magnetic systems (corresponding to a modulation of the local magnetization) and in intercalated laminar compounds.

The elementary excitations in incommensurate phases have specific peculiarities. In the framework of quasi-harmonic theories for lattice vibrations it has been shown that there are two fluctuating coordinates: the first corresponds to a spatial modulation of the phase of the displacement wave (phasons); the second corresponds to the fluctuations in the amplitude of the wave (amplitudons). In some circumstances the spatial modulations can be highly non-linear. In this case the incommensurate phase can be viewed as a succession of commensurate regions separated by regions where the "local order parameter" varies rapidly. The Hamiltonian for such a configuration can be cast in the Sine-Gordon form in the continuum classical limit. Thus the dynamics of this region of discommensuration can be described in terms of domain walls or phase solitons. Furthermore, the lock-in transition to the commensurate phase can be described in terms of a vanishing density of phase solitons [2.14, 33] (Panel 2.7).

2.2 Basic Concepts and Methodology of NMR–NQR in Crystals

NMR–NQR spectroscopy conveys simultaneously information on characteristic properties of the nuclei (spin, magnetic and quadrupolar moments) and on properties of the "lattice", namely the local environment surrounding the nuclei [2.34, 35]. Since the nuclear properties are by now well known, the nuclei become non-perturbing microscopic probes through which a variety of information on static and dynamical properties of the crystal can be deduced. NMR spectroscopy is based on the generation of nuclear Zeeman levels by an external magnetic field H_0. In NQR (nuclear quadrupole resonance) no external field is necessary because the degeneracy of the nuclear levels (in the magnetic quantum number) is partially removed by the electric field gradient (EFG), which couples to the quadrupole moment eQ of the nucleus. When both Zeeman splitting and quadrupole splitting are present, one has, in general, a complicated structure of levels and phenomena, which simplifies in the two limiting cases where one of the two interactiions can be treated as a small perturbation of the other (Panels 2.8 and 2.9).

Incommensurate Phase Transitions

- At the incommensurate PT the atomic displacement or the statistical average of the order-disorder variable has a spatial periodicity incommensurate with respect to the underlying lattice. In the plane-wave model, the excitations are phase waves and amplitude waves (Fig. 2.7a). In the soliton-like model there are regions of discommensuration, separating commensurate domains, which propagate like solitons (Fig. 2.7b).

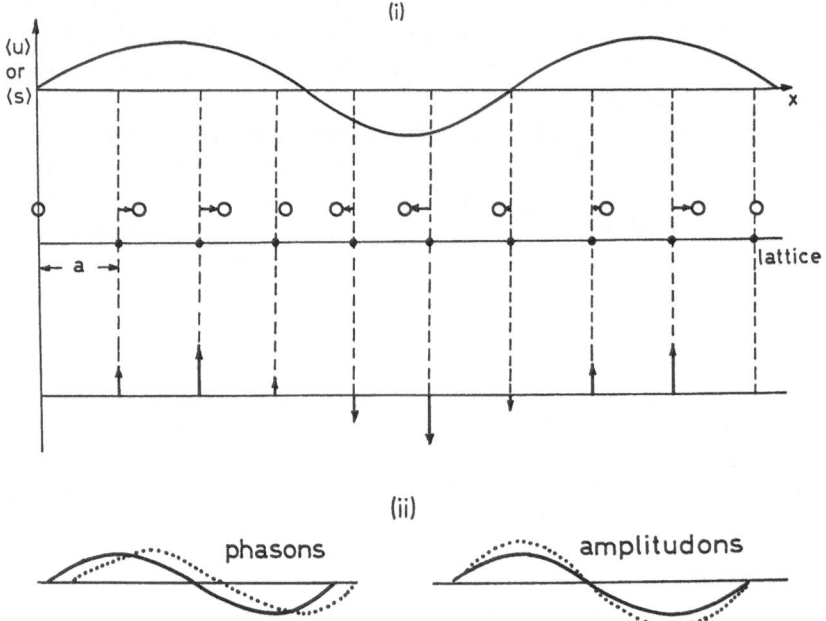

Fig. 2.7a. Atomic displacements $\langle u \rangle$ or statistical value $\langle s \rangle$ along one direction in an incommensurate phase (i) and sketch of the wave for excitations of phason and amplitudon character (ii)

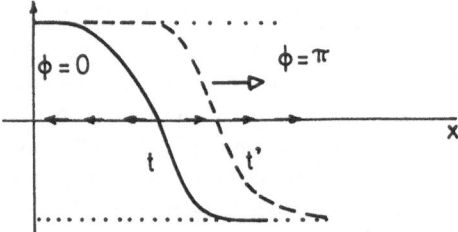

Fig. 2.7b. Illustrative representation of the propagation along the x-direction of the region of discommensuration in the soliton-like model of the incommensurate phase

NMR–NQR Hamiltonians and Eigenstates

- The interaction of the nuclear moments $\mu = g\mu_N I = \gamma\hbar I$ (g, Landé factor of the order of unity, μ_N nuclear magneton, γ gyromagnetic ratio) with the external field H_0 generates the Zeeman levels among which NMR spectroscopy is carried out by inducing $\Delta m = \pm 1$ magnetic dipole transitions with rf photons having circular polarization. The Hamiltonian is

$$\mathcal{H}_z = -\gamma\hbar H_0 \sum_i I_z^{(i)}$$

and for half-integer spin $I^{(i)}$ the energy levels are shown in Fig. 2.8a.

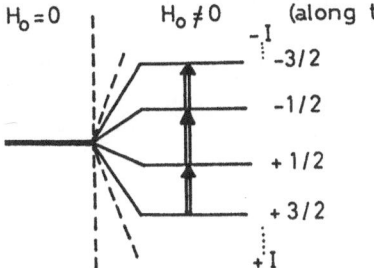

Fig. 2.8a. NMR Zeeman levels generated by the external field H_0 for half-integer nuclear spin

- The interaction of the nuclear quadrupole moment eQ with the electric field gradient (EFG) partially removes the degeneracy in the quantum number m. NQR spectroscopy involves dipole moment transitions among these levels. The Hamiltonian is

$$\mathcal{H}_Q = \sum_i \mathcal{H}_Q^{(i)} = \sum_i \frac{e^2 qQ}{4I(2I-1)}\left[3I_z^2 - I(I+1) + \frac{1}{2}\eta\left(I_+^2 - I_-^2\right)\right]$$

(where $eq = V_{zz} = \partial^2 V/\partial z^2$ and V_{JK} are the EFG components in the principal axes frame of reference). The NQR levels, for the case of an asymmetry parameter $\eta = |V_{XX} - V_{YY}|/V_{ZZ}$ equal to zero, have the structure depicted in Fig. 2.8b.

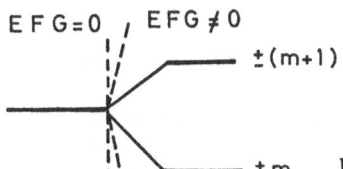

Fig. 2.8b. NQR levels ($H_0 = 0$ and asymmetry parameter $\eta = 0$)

NMR–NQR Experiment: Signal Generation and Detection

- The resonance transitions are induced by a rf field H_1 along a coil axis perpendicular to the external field H_0 (in NMR) or to V_{zz} in (NQR). The rf field corresponds to a small rotating magnetic field. At resonance the precessional frequency is equal to ω_{rf} (Fig. 2.9a).

- In a frame $x'y'z'$ rotating at ω_{rf} (Fig. 2.9a) one has, at resonance, the precession of the magnetization around H_1, with an angular frequency $\omega_1 = \gamma H_1$ (as if H_0 were absent) (Fig. 2.9b). For a duration of the rf pulse τ the magnetization rotates by an angle $\theta = \gamma H_1 \tau$.

  By controlling the duration of the rf pulse, the nuclear magnetization can be prepared along any direction. $\theta = \pi/2$ corresponds to the saturation of two levels in the quantum mechanical description of transitions induced by the rf photons; $\theta = \pi$ corresponds to the inversion of the populations among the levels involved in the irradiation. These non-linear effects require a strong rf field.

- The NMR or NQR signal $s(t)$ follows the rf pulse, being generated by the free precession of the nuclear moments in the xy plane, at the end of a rf pulse of length $t = (\pi/2)/\gamma H_1$ where $2H_1$ is the amplitude of the rf field (Fig. 2.9c).

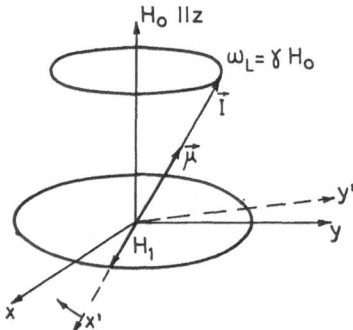

Fig. 2.9a. The resonance of the precessional Larmor freqency $\omega_L = \gamma H_0$ (in NMR) with the rotational frequency of H_1 in the xy plane

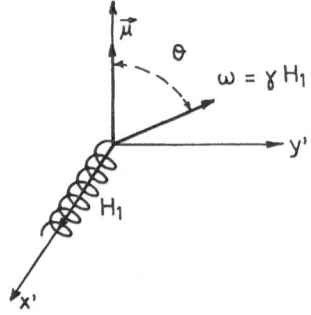

Fig. 2.9b. Motion of an individual nuclear magnetic moment at the resonance, as seen in the rotating frame

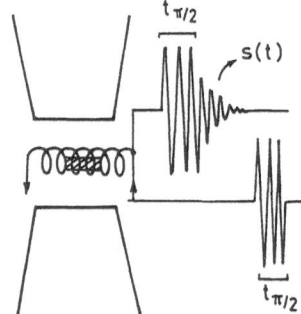

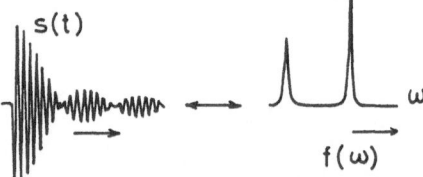

Fig. 2.9c. Illustration of the detection of the signal $s(t)$ following the rf pulse (free induction decay – FID) and of the Fourier transformation (FT) yielding the corresponding spectrum in the frequency domain

NMR–NQR spectroscopy typically involves the frequency range 1–300 MHz and is characterized by the following operational features: (i) a large number of photons per unit energy; (ii) strong coherence of the radiation; (iii) a very low rate of spontaneous emission; (iv) a low rate of relaxation processes. These features allow one to carry out unique experiments (in the spectroscopy field) in which the system of the nuclei can be prepared in given initial states and its time evolution can be monitored continuously in time.

Recent techniques utilize almost exclusively pulse methods with Fourier transformation (FT): a very narrow rf pulse irradiates the sample, its response in the time domain is collected (and, if necessary, accumulated to enhance the signal-to-noise ratio) and the Fourier inversion yields the spectrum in the frequency domain. Moreover, this technique is particularly suitable for the measurement of relaxation times.

2.2.1 NMR–NQR Rigid-Lattice Spectra

When motional effects can be neglected, the resonance spectra can be interpreted directly in terms of the static parts of the interactions among nuclei and between the nuclei and the surrounding lattice.

In a non-metallic diamagnetic crystal, with resonant nuclei having spin $I = \frac{1}{2}$ (no quadrupole moment), one normally observes a single resonance line, centered at the Larmor frequency $\nu_L = \gamma H_0$, whose shape and width reflect the nuclear dipole-dipole interactions.

The dipolar broadening is usually of the order of a few Gauss corresponding to a frequency width $\delta\nu$ which ranges from around 10 kHz for protons to about 1 kHz for nuclei with a small magnetic moment like ^{89}Y.

A line width $\delta\nu$ corresponds to a time decay constant T_2 for a Gaussian free precession decay given by $T_2 \sim \frac{1}{2}\delta\nu^{-1}$. Thus in solids it is important to reduce as much as possible the dead time of the receiver that necessarily follows the application of a strong rf pulse.

The diamagnetic currents of the electron clouds cause a small correction to the external field H_0, inducing a shift in the resonance frequency which depends on the chemical environment (*chemical shift*). The order of magnitude of the shift is usually 10–100 ppm. In order to resolve the resonance lines of nuclei in non-equivalent sites, one has to average out the dipolar broadening by resorting to special pulse sequences or to magic-angle spinning. When chemical shifts can be detected, their measurement directly probes the modifications occurring at a SPT.

In metallic systems the hyperfine magnetic field due to conduction electrons induces a resonance shift (Knight-shift) of the order of 1–0.1 %. Knight-shift measurements yield information about Pauli susceptibility and about the character of the conduction electron wave functions at the Fermi level.

In paramagnetic substances, the magnetic interaction of the nuclei with the unpaired electron moments induces shifts of resonance frequency of the order of a few percent. The measurement of the paramagnetic shift yields important information about the static susceptibility and the electronic wave functions. Panel 2.10 illustrates the broadening and the different types of shift.

Rigid-Lattice NMR Spectra

- The distribution of the internal fields due to the nuclear dipole-dipole interaction broadens the resonance line symmetrically around the Larmor frequency ω_L. The dipole-dipole interaction is described by the Hamiltonian

$$\mathcal{H}_{dip}^{1,2} = \gamma_1 \gamma_2\, \hbar^2 I_1 \cdot \left\{ (I_2/r_{12}^3) - 3\left[r_{12}(I_2 \cdot r_{12})/r_{12}^5 \right] \right\} \quad .$$

  The width of the resonance spectrum (usually gaussian) $f(\omega - \omega_L)$ (Fig. 2.10a) is $\delta = 2.36\Delta$ where the second moment Δ^2 is given by

$$\Delta^2 \equiv \int \omega^2 f(\omega - \omega_L)\, d\omega = \tfrac{3}{4}\gamma^4 \hbar^2 I(I+1) \sum_k \left(1 - 3\cos^2\theta_{1k}\right)^2 / r_{1k}^6$$

  (for like spins).
- The diamagnetic shielding of the electronic currents causes a shift of the resonance line, and therefore a splitting of the lines in the case of non-equivalent resonant nuclei (Fig. 2.10b).
- The Knight shift of the NMR line in metals is caused by the hyperfine magnetic interaction with the conduction electrons (Fig. 2.10c).
- The "paramagnetic shift" of an NMR line is due to the hyperfine interaction with unpaired electron spins (Fig. 2.10d).

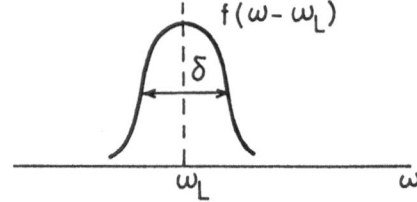

Fig. 2.10a. Broadening of a resonance line due to the dipole-dipole interaction

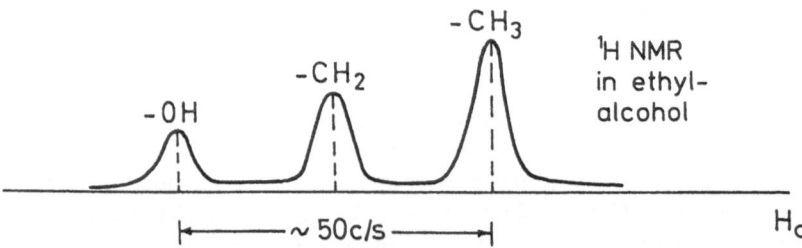

Fig. 2.10b. Example of an NMR spectrum (with poor resolution) displaying the effect of the chemical shift. (The rf frequency is $\nu_L = 30\,\text{MHz}$ and $H_L \equiv \omega_L/\gamma$)

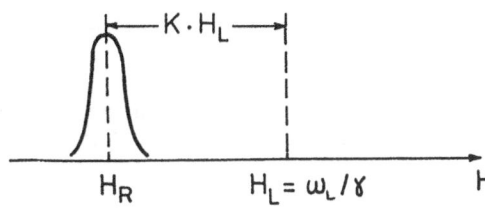

Fig. 2.10c.
Example of Knight shift in the ^{27}Al NMR line in aluminum metal (powder). The Knight shift is defined as $K = (H_L - H_R)/H_L \cdot 100$ and is about 0.16 % in this case

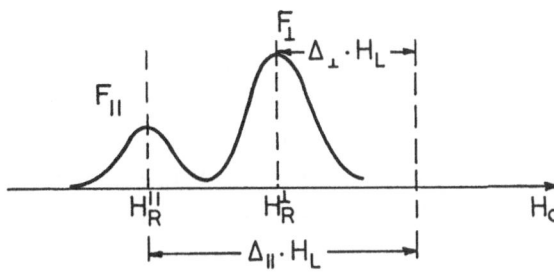

Fig. 2.10d. Example of paramagnetic shift in the ^{19}F NMR line in a single crystal of RbMnF$_3$. The paramagnetic shift, Δ, is defined like the Knight shift. ($\parallel$ and $\perp$ refer to the direction of H_0 with respect to the MnF bond.) One has $\Delta_\parallel = 2.95$ % and $\Delta_\perp = 1.8$ %

When there is a distribution of local hyperfine fields and/or dipolar interactions the spectrum is broadened, sometimes asymmetrically, and can develop a structure. In this case there is a correspondence between the frequency distribution of the spectrum and the spatial distribution of the local fields in the sample. This circumstance is particularly useful when investigating phase transitions in crystals with paramagnetic ions or impurities and CDW transitions.

In studying SPTs and disordering phenomena it is particularly useful to use as a microscopic probe a nucleus with I greater than $\frac{1}{2}$, possessing a nuclear quadrupole moment eQ. In the NMR rigid-lattice spectra the quadrupole interaction causes splittings of the Zeeman levels. In a single crystal and for half integral I, the quadrupole-perturbed NMR spectrum consists of a central line at $\sim \nu_L$ and $(2I - 1)$ satellite lines symmetrically located. The EFG tensor at the nuclear site can be obtained from the angular dependence of the spectra.

For a strong enough quadrupole interaction a detectable shift of the central line $\Delta\nu_{\frac{1}{2}}$, with respect to ν_L, occurs. This can be calculated by considering the second order effects of the quadrupole interaction on the Zeeman levels. Often, due to imperfections and/or strain fields around dislocations or impurities, the satellite lines are broadened because of the distribution in magnitude and orientation of EFGs. Then the less sensitive second-order quadrupole shift still affords an easy way to determine the EFG. Such a situation is commonly found in powdered samples, due to the random orientation of the crystallites with respect to H_0 (Panel 2.11).

In studies of SPTs the analysis of the quadrupole-perturbed NMR spectra has given valuable information. In particular, transitions undetected by other techniques have been discovered and the structure and the local symmetry of the low temperature phase have been obtained. Moreover, the temperature dependence of the order parameter can be derived from the EFG. The sensitivity is better, in certain cases,

Effects of the Quadrupole Interaction

- The NMR spectrum resulting from the perturbation due to the quadrupole inter-action is easily derived up to second order. The effects are depicted in Fig. 2.11a, where

$$\Delta \nu_m^{(1)} = \nu_Q(2m-1)\left(3\cos^2\theta - 1 + \eta \sin^2\theta \cos 2\varphi\right)/4$$

$$\Delta \nu_{1/2}^{(2)} = -\left(\nu_Q^2/16\nu_L\right)\left(a - \tfrac{3}{4}\right)\left(1 - \cos^2\theta\right)\left(9\cos^2\theta - 1\right)$$

$$\text{for}\quad \eta = 0$$

with $a = I(I+1)$ and $\nu_Q = 3e^2qQ/2hI(2I-1)$.

- The NQR spectrum (with no external magnetic field H_0) cannot easily be eval-uated in the case of non axially symmetric EFGs. For $\eta = 0$ (and half integer spins) one has $(2I-1)/2$ lines separated by ν_Q, as depicted in Fig. 2.11b.

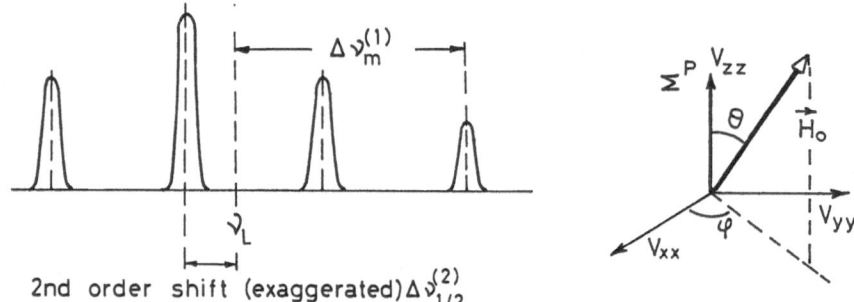

Fig. 2.11a. Sketch of the quadrupole-perturbed NMR spectrum (Σ^P: frame of reference of the principal axes of the EFG tensor)

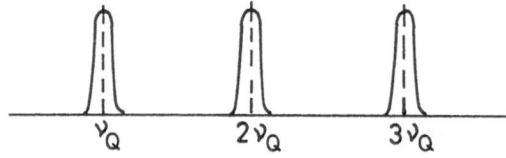

Fig. 2.11b. Sketch of the NQR spectrum with no magnetic perturbation, half inte-ger spin, and symmetric EFG tensor

than that of X-ray diffraction and atomic displacements of the order of 0.01 Å can be detected.

For nuclei having a large quadrupole moment in covalently bonded crystals, as is often the case for Cl, Br, Cu and others, the strength of the quadrupole interaction reaches the MHz or 10 MHz range. The external magnetic field can be removed and the rf field can induce transitions directly between the eigenstates of the quadrupole Hamiltonian. The quantization axis is now the Z principal axis of the EFG ten-sor with greatest component V_{ZZ}, so the maximum signal is obtained when H_1 is perpendicular to Z, with no orientational dependence of the NQR frequencies.

However, the orientation of the EFG tensor in the crystal can still be obtained, by applying a small external magnetic field against which the crystal is rotated.

2.2.2 Motional Effects on the Spectra

Through the time dependence of the interactions the motions drastically affect the rigid-lattice NMR–NQR spectra.

For NMR spectra, a rather general rule that should be emphasized at the start is the following. When the characteristic frequencies of the motions modulating the interactions are of the same order of magnitude as the frequency strengths of the interactions themselves, then one has a narrowing of the spectra, namely broad rigid-lattice lines become narrow, while multicomponent structures collapse into a single line.

One can relate the line-shape of the resonance spectrum to the characteristics of the microscopic motions by using a time-dependent approach and linear response theory. The most common limiting situations are a Gaussian line for the rigid-lattice condition, which turns into a narrow Lorentzian line for the case of fast motions (Panel 2.12).

The FID signal $s(t)$ gives the relaxation function $\varphi(t)$ directly, while the spectrum is its FT centered at the resonance frequency. The details of the motions are reflected in the relaxation function and consequently change the resonance spectra. The approach outlined above can also be used when the motion modulating the interaction is an atomic jump between non-equivalent sites. The effect on the relaxation function is to cause the multicomponent structure of the line to collapse.

The motional effects on NQR spectra differ in important aspects from the ones of pure or quadrupole-perturbed NMR. The main part of the NQR Hamiltonian often arises from the coupling with the local EFG: the motions directly affect the eigenstates thus causing variations of the energy spacing of the levels. As a consequence the NQR resonance frequency is, in general, rather strongly temperature dependent. Furthermore, the motions generally cause line broadening with maximum effectiveness when their frequencies are of the order of the quadrupole resonance frequency.

When the correlation times characteristic of the motions become longer than the inverse resonance frequencies, only rigid-lattice spectra of the same origin as in NMR spectra are present. If the motion involves jumps between positions where the nucleus experiences a different EFG, a distinct spectrum is obtained reflecting each of the sites (Panel 2.12).

2.2.3 Relaxation Processes and Relaxation Rates

The time-dependent part of the interactions between the nuclei and the surrounding lattice drives the relaxation process, namely the process which causes the spin system to attain its equilibrium distribution. Measurement of the relaxation rate yields valuable information about the microscopic motions. Typically, the relaxation rate is measured by intentionally altering the equilibrium distribution among the Zeeman

Motional Effects on NMR–NQR Spectra

- By considering the effect of a time-dependent interaction in a perturbative, linear-response approach, one can prove that the line-shape is the Fourier transform of the correlation function for magnetization perpendicular to H_0:

$$f(\omega - \omega_0) \propto \int e^{-i(\omega-\omega_0)t} \langle M_x(t)M_x(0)\rangle \, dt \quad ,$$

where $\langle M_x(t)M_x(0)\rangle = \varphi(t)$ is the relaxation function.

  The spread in the resonance frequencies $\Delta^2 = \langle \omega^2(0)\rangle$, corresponding to the $t = 0$ correlation function for the motion, is averaged out by fast motions when the correlation frequency $\omega_c \gg (\Delta^2)^{1/2}$ (Fig. 2.12a).

- When the motion consists in the jumping of an atom between inequivalent sites, the collapse of the two "rigid-lattice" lines into a single line occurs when the frequency of the motion is larger than the difference between the two resonance frequencies of the individual line (Fig. 2.12b).

- In NQR, the motions in some cases move the Σ^P frame of the reference of the EFG tensor with respect to a crystalline frame of reference Σ^C (Fig. 2.12c), causing effects which differ from those in NMR.

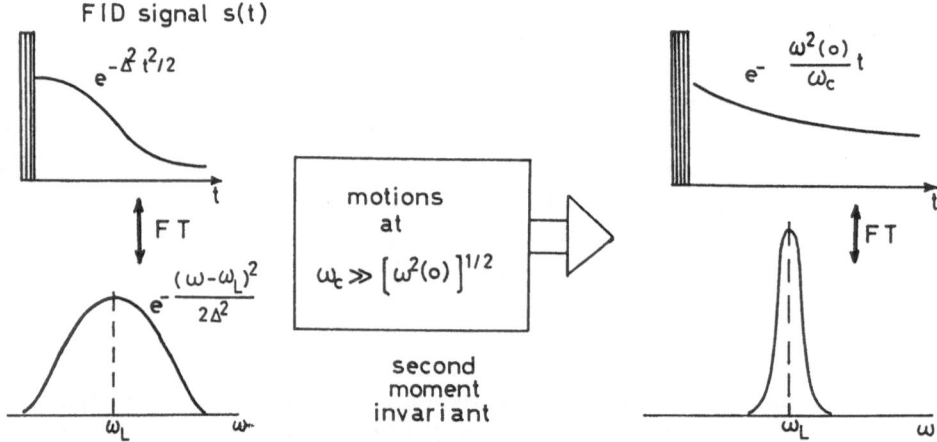

Fig. 2.12a. Illustration of the effect of fast motions in causing the FID to decay in a time $\tau = \omega_c/\omega^2(0)$, much longer than that in the rigid-lattice situation and, correspondingly, in causing the narrowing of the resonance line

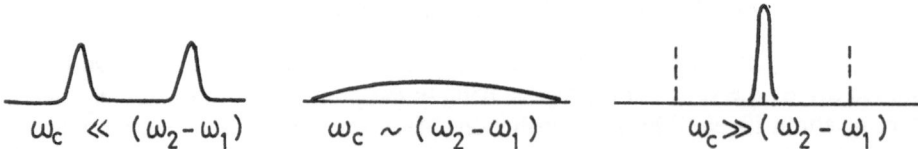

Fig. 2.12b. Illustrative sketch of the collapsing of resonance lines by the jumping of an atom between inequivalent sites

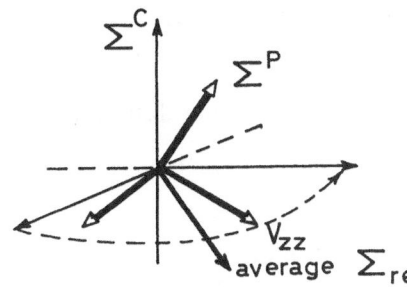

Fig. 2.12c. Motion of the Σ^P frame of reference of V_{JK}: the effective quantization axis is along an average reference axis; the resonance frequency is proportional to the average $\langle V_{JK}^C \rangle$ while the line-width (and the relaxation times) are due to the fluctuating part $V_{JK}^C(t) - \langle V_{JK}^C \rangle$

energy levels and subsequently monitoring the recovery towards equilibrium of the nuclear magnetization. There are several methods of detecting relaxation processes. The simplest, at least conceptually, is to apply a short rf pulse of strength and duration such that all the Zeeman levels become equally populated (corresponding to an infinite spin temperature) and then through a magnetic resonance signal to record the growth in time of the population difference between two adjacent levels.

One can define different relaxation times that are differently related to the features of the microscopic motions. Here we only define T_1, $T_{1\varrho}$ and T_2 and describe how they can be measured. T_1, the spin-lattice relaxation time, measures the recovery rate of the component of the nuclear magnetization along the z-direction of the external magnetic field in NMR. In NQR, the description in terms of nuclear magnetization is inappropriate, yet the resonance signal at time t after the saturating pulse is proportional to the populations of the energy levels, thus allowing the measurement of a relaxation rate.

$T_{1\varrho}$ is the spin-lattice relaxation time for the nuclear magnetization in the presence of a strong rf field, whose phase is properly adjusted to lock the magnetization along the rf field (so that no transitions are induced) [2.36]. In practice, $T_{1\varrho}$ is measured, in NMR, by applying a saturating rf pulse immediately followed by a pulse dephased by $\pi/2$; the signal detected at the end of this second locking pulse decays as a function of the pulse duration, with time constant $T_{1\varrho}$.

T_2, the spin-spin relaxation time (or dephasing time of the free induction decay) is the time in which the transverse nuclear magnetization, as measured by the free precession decay, falls to $1/e$ of its initial value following a $\pi/2$ pulse. The decay of the transverse nuclear magnetization can be caused by different mechanisms. When the dominant mechanism is an inhomogeneous distribution of resonance frequencies (as due for instance to an inhomogeneous magnetic field or to a distribution of static quadrupole interactions or paramagnetic shifts), the dephasing is of *extrinsic* origin and one normally speaks of T_2^*. In such a case the dephasing time associated with the intrinsic interaction mechanism (normally dipole-dipole interaction) can be measured from the decay time of the amplitudes of the echo signal which can be generated by appropriate pulse sequences (Panel 2.13). For details on the experimental methods see Ref. [2.35].

The experimentally measured T_1, T_2 and $T_{1\varrho}$ yield information about the microscopic dynamics which modulate the nucleus-lattice interactions or cause energy

Relaxation Times: Definition and Measurement

- By referring to an $I = \frac{1}{2}$ nucleus in NMR (Fig. 2.13a), the equilibrium magnetization is

$$M_Z(\infty) \propto \left(N_{1/2} - N_{-1/2}\right) \propto \hbar\omega_{\mathrm{L}}/k_{\mathrm{B}}T \quad .$$

After a saturating $\pi/2$ rf pulse the transient magnetization is

$$M_Z(t) = M_Z(\infty)\left[1 - \exp\left(-t/T_1\right)\right] \quad ,$$

and the spin-lattice relaxation time T_1 can be measured by applying a second $\pi/2$ rf pulse (Fig. 2.13b).
- The spin-lattice relaxation time for the nuclear magnetization aligned along H_1, $T_{1\varrho}$, can be measured by monitoring the "output" signal at the end of the "locking" pulse (Fig. 2.13c). During the time the rf is on, the magnetization decays to its equilibrium value in H_1 ($H_1 \ll H_0$) staying collinear with H_1.
- The spin-spin relaxation time T_2 is measured directly from the FID signal following a $\pi/2$ pulse. When the dephasing of the magnetization in the xy plane is due to inhomogeneous broadening (yielding the so-called T_2^*), then T_2 can be obtained with the echo technique (Fig. 2.13d).

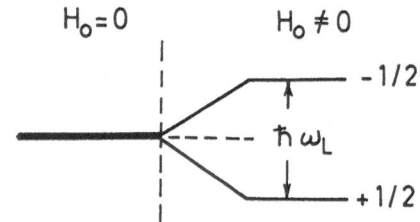

Fig. 2.13a. Zeeman NMR levels for a spin one-half nucleus

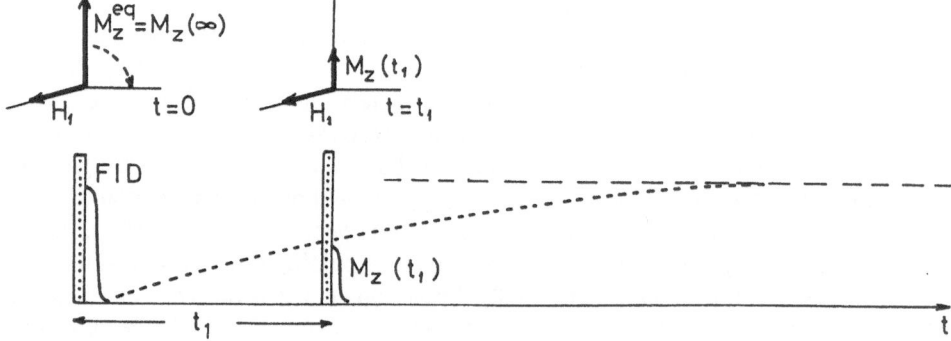

Fig. 2.13b. The measurement of the spin-lattice relaxation time T_1 with two $\pi/2$ rf pulses

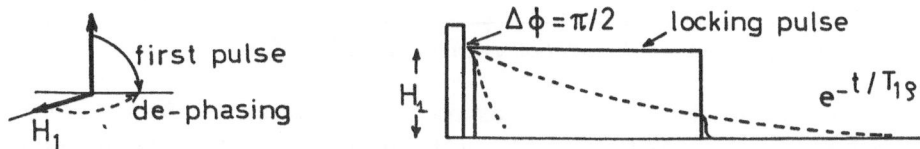

Fig. 2.13c. The measurement of the spin-lattice relaxation time in the rotating frame with a $\pi/2$ rf pulse and a second rf pulse of duration t applied immediately after the first, with a dephasing of $\pi/2$

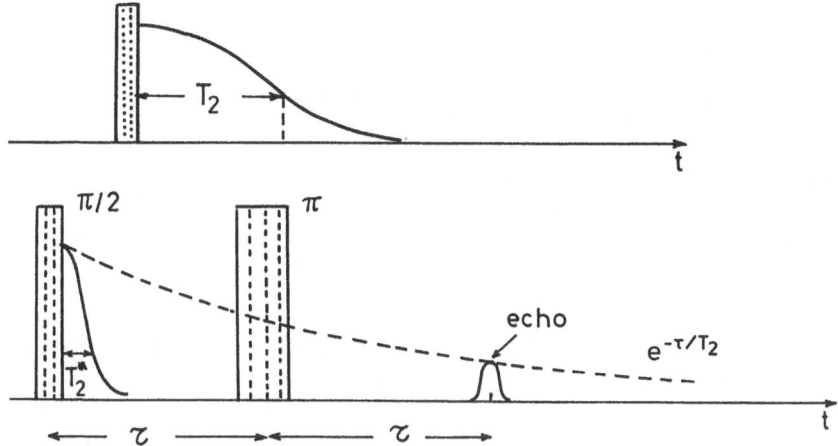

Fig. 2.13d. T_2 is the decay time of the FID signal (see also Panel 2.12). In the presence of inhomogeneous broadening of the resonance line, T_2 can be measured from the amplitude of the echo signal following a second rf pulse

exchange between nuclei and the elementary excitations of the lattice. For this purpose it is crucial to have reliable theories to relate the relaxation rate to the lattice functions and their dynamics [2.37, 38]. It is customary to refer to two limiting model situations: (i) *quantum lattice*; (ii) *classical lattice* (Panel 2.14).

In the first case one considers the quantum states of the entire system, nucleus plus lattice, and defines the relaxation transition probabilities in terms of matrix elements of the interaction Hamiltonian among the quantum states. This approach can be used, for example, to describe nuclear relaxation in metals induced by the hyperfine interactions with the conduction electrons and nuclear quadrupole relaxation in crystals induced by the lattice vibrations. One can define a *direct process* in which a nuclear spin makes a transition $|m\rangle \rightarrow |m'\rangle$ exchanging an energy $\hbar\omega_{mm'} \sim \hbar\omega_L$ with the lattice excitations. The *indirect process* proceeds via a nuclear spin transition accompanied by absorption and emission of a pair of excitations whose energies differ by $\hbar\omega_L$.

When the fluctuations of the lattice interactions are due to random, or largely incoherent motions, such as diffusion processes or strongly anharmonic lattice vibrations with short life times, then it is more appropriate to refer to a classical-lattice mode. In this case, only the nuclear spin system is quantal while the interactions with

The Relaxation Process and Relaxation Transition Probabilities

- To relate the relaxation times to the microscopic dynamics one must refer to the relaxation transition probabilities $W_{m,m'}$, for quantum and classical lattices, in the weak-collision, perturbative approach:

Quantum Lattice

$$W_{mm'} = \frac{2\pi}{\hbar} \sum_{ff'} P_f |\langle fm|\mathcal{H}'|f'm'\rangle|^2$$
$$\times \delta\big(E_f + E_m - E_{f'} - E_{m'}\big)$$

where $|fm\rangle$ and $|fm'\rangle$ are the initial and final states, and $\mathcal{H}'$ is the perturbing time dependent Hamiltonian.

The quantum lattice model is typically applied for quadrupole NMR relaxation due to the lattice vibrations:

Fig. 2.14

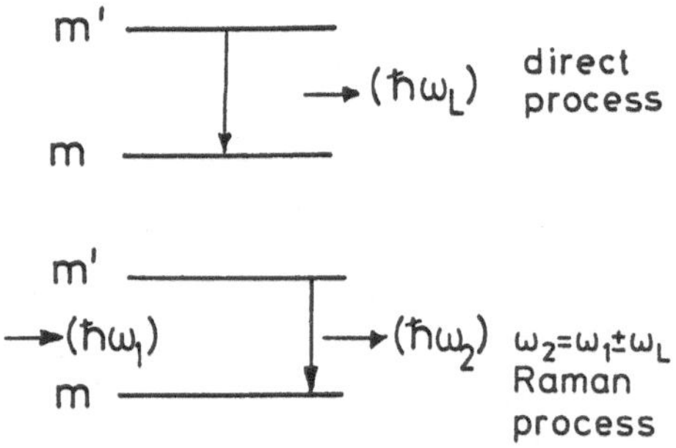

Usually the Raman process dominates and one has

$T_1^{-1} \propto T^2$ for $T \gtrsim \theta_{\text{Debye}}$
$T_1^{-1} \propto T^7$ for very low temperatures, with no ω_L-dependence.

Classical Lattice

$$W_{mm'} = \frac{1}{\hbar^2} \int e^{-i\omega_{mm'}\tau}$$
$$\times \langle \mathcal{H}'_{mm'}(0)\mathcal{H}'_{mm'}(\tau)\rangle_0 \, d\tau$$
$$\equiv \frac{|\mathcal{H}'_{mm'}(0)|^2}{\hbar^2} J(\omega)$$

where $\langle \ \rangle$ is the statistical ensemble average and $J(\omega)$ is the spectral density for the correlation function of the variables coupling the nuclear spins and the lattice.

The classical-lattice model is normally applied for diffusion-like processes, for short-lived or anharmonic phonons, or in fluids.

the lattice are random functions of time. One then uses time-dependent perturbation theory to express the relaxation transition probabilities $W_{mm'}$. This semiclassical approach can be carried out in the so-called "weak-collision" or "strong-collision" limits. The weak-collision limit is applicable when many fluctuation events are necessary to induce a sizeable relaxation probability (i.e. many events occur in a time of the order of T_1). Thus the W's can be related to the spectral densities of the motions. The strong collision approach should be used when each fluctuation event is capable of inducing a sizeable modification of the interactions and/or of the direction of the eigenvectors (for example in NQR when the EFG reorients because of slow motions, or in NMR in paramagnets when the fluctuating hyperfine magnetic field is very strong). In the strong-collision limit the relaxation time which is measured is directly related to the time characterizing the fluctuation of the interaction.

The general features of the relaxation processes can be obtained by referring to a simple random motion of relaxation type, which modulates the interaction Hamiltonian. In this case the correlation function is approximately described by an exponential decay, with a single correlation time τ. The correlation time τ is normally a function of temperature and the behavior of the relaxation rates is shown schematically in Panel 2.15.

PANEL 2.15 _____

The Relaxation Process and Spectral Densities for Random Motions

- For random motions of a simple relaxational type the correlation function for $\mathcal{H}'(t)$ is $\varrho(t) = |\mathcal{H}'(0)|^2 \exp(-t/\tau_c)$ and the spectral density is $J(\omega) = 2\tau_c/(1 + \omega^2\tau_c^2)$.

  The general behavior of the relaxation rate can be discussed in terms of three regimes:

  a) Fast motion, namely $\omega_c = \tau_c^{-1} \gg \omega_L$; one then has, for the spectral density depicted in Fig. 2.15a,

  $$J(\omega_L) \simeq J(\omega_1) \simeq J(0) = 2\tau_c \quad ,$$

  thus yielding

  $$T_1^{-1} \simeq T_2^{-1} \simeq T_{1\varrho}^{-1} = \frac{|\mathcal{H}'|^2}{\hbar^2}2\tau_c$$

  independent of ω_L.

  b) Intermediate motions, namely characteristic frequency ω_c around ω_L (Fig. 2.15b); in this case one has a highly effective spin-lattice relaxation mechanism and

  $$T_1^{-1} \simeq \frac{|\mathcal{H}'|^2}{\hbar^2}\frac{1}{\omega_L} \quad .$$

  c) Slow motion regime, namely $\omega_c \ll \omega_L, \omega_1$; then

  $$T_1 \neq T_{1\varrho} \neq T_2 \quad \text{and}$$
  $$T_1^{-1} \sim \frac{|\mathcal{H}'|^2}{\hbar^2}\left(\omega_L^2\tau_c\right)^{-1}$$
  $$T_2^{-1} \sim \text{linewidth}$$
  $$T_{1\varrho}^{-1} \simeq \frac{|\mathcal{H}'|^2}{\hbar^2}\left(\omega_1^2\tau_c\right)^{-1} \quad .$$

  The typical behavior of the relaxation rates due to random motions as a function of the correlation time is shown in Fig. 2.15c.

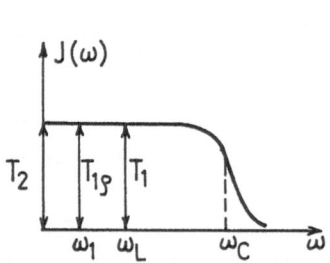

Fig. 2.15a. Spectral density and relaxation rates for random motions in the fast-motion regime

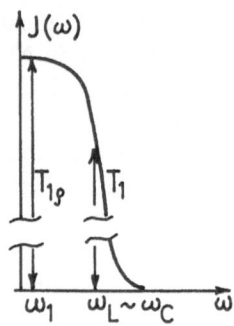

Fig. 2.15b. The situation of motion with a characteristic frequency around ω_L

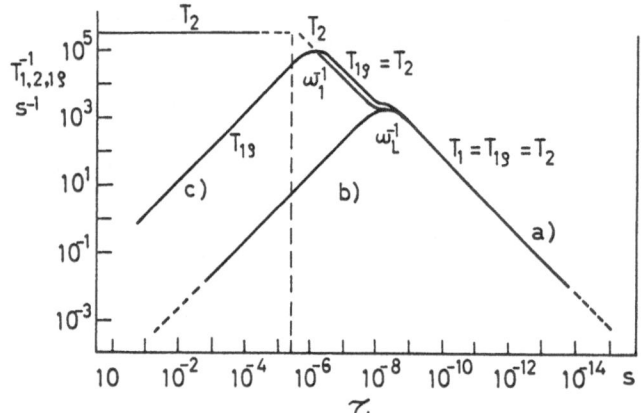

Fig. 2.15c. The dependence of the relaxation rates on the correlation time of random motions, for a given resonance frequency ω_L and a given strength of the interaction

2.3 Phase Transitions and Critical Effects in Ordered Crystals: Insights from NMR–NQR

An exhaustive review of NMR–NQR studies of SPTs, discussing the many aspects of phase transitions and critical phenomena where NMR–NQR have provided important insights, has already been published [2.38]. In this section we present a few illustrative examples of investigations representative of the type of information that can be gained by NMR–NQR in "ordered" materials. All aspects related to disorder, non-linearity and other "non-conventional" effects will be the object of detailed treatment in subsequent sections.

2.3.1 Order Parameter Studies

The study of quadrupole-perturbative effects on NMR spectra offered a powerful tool to look into the static phenomena occurring at the antiferro-distortive transitions involving rotations of the BX_6 octahedra in perovskite crystals ABX_3. In these cases second-order or quasi-second-order SPTs occur, whereby the order parameter corresponds to the time-averaged rotation angle $\langle \varphi \rangle$ in a cell. The determination of the critical exponent β is of paramount importance for the concept of universality in phase transitions [2.39].

In the low temperature phase, the rotations of the BX_6 octahedra cause the onset of an average EFG component $\langle V_{zz} \rangle$ at a given nucleus. $\langle V_{zz} \rangle$ can be related to the generalized order parameter by symmetry arguments. Furthermore the relationship of $\langle V_{zz} \rangle$ to φ will depend on the nuclear site (A, B or X). Thus the temperature dependence of $\langle V_{zz} \rangle$, as deduced either from the first-order splitting of the satellite line, or from the second-order shift of the central line, directly yields the temperature dependence of the order parameter.

In KMnF$_3$, through ^{39}K NMR, the rotation angle has been shown [2.40] to display a Landau-type behavior with $\beta = \frac{1}{2}$ over a large temperature range, with a renormalization of the limiting temperature T_0, for $T \gtrsim T_c - 20$. Close to T_c a critical exponent $\beta \sim \frac{1}{3}$ was obtained by analyzing the data in terms of a second-order transition. The first-order character of the transition (a discontinuity of about $3°$ has been deduced from birefringence measurements) makes an analysis of the data uncertain in terms of the critical exponent close to T_c.

In RbCaF$_3$ the analogous transition is almost second order (the discontinuity in the birefringence is small), and a more reliable estimate of the β exponent in the critical region can be obtained from the temperature dependence of the second-order quadrupole shift (Panel 2.16). The strict resolution limit ($\Delta \gtrsim \delta/2$) can often be somewhat improved by an appropriate fitting procedure. In RbCaF$_3$, where $\Delta \propto \varphi^2$, even using the stricter resolution condition it was possible [2.41] to follow the rotation angle close enough to T_c to determine without ambiguity a critical exponent $\beta \simeq 0.27$ for $\varepsilon \lesssim 5 \times 10^{-1}$, undoubtedly less than the value $\beta = \frac{1}{3}$ required for the 3D Heisenberg universality class.

In NaNbO$_3$ (Panel 2.16) the concomitant effects of $\Delta \propto \varphi$ (rather than φ^2) and the high transition temperature, made an investigation with much higher resolution possible. A value of the critical exponent $\beta = 0.17 \pm 0.02$ was derived [2.42]. Furthermore, with a detailed fitting procedure of the spectra and from the gradual disappearance, below T_c, of the satellite lines, the second-order character and of the transition was established.

A nice example of a study of the temperature dependence of the order parameter from NQR spectra, rather than from quadrupole-perturbed NMR spectra, is offered by the ^{35}Cl NQR study of K$_2$ReCl$_6$ [2.43]. The EFG at the Cl site has V_{zz} directed along the Re-Cl bond and is axially symmetric. An external magnetic field H_0 induces a Zeeman perturbation on the NQR levels and by applying it along one of the cubic axes it is possible to follow the behavior of one of the two NQR lines relative to those nuclei having the Re-Cl bond direction along the field. Below T_c, as a consequence of the rotation of the octahedra, for two-thirds of Cl nuclei the bond direction is no longer along H_0 and the observed resonance line splits into two components. The angle of rotation of the ReCl$_6$ octahedra is evaluated by rotating the crystal in the magnetic field (Panel 2.17).

An example of a very precise, although rather indirect, determination of the order parameter is offered by the ^{35}Cl NQR study in (CH$_3$NH$_3$)$_2$MnCl$_4$ around the

NMR Studies of the Order Parameter

- For the SPT in perovskite, the freezing of a particular soft mode corresponds to rotations of the octahedra in the same (M_3 mode) or in opposite sense (R_{25} mode) in adjacent planes (Fig. 2.16a).
- The NMR line corresponding to the $+\frac{1}{2} \leftrightarrow -\frac{1}{2}$ transition is shifted by static quadrupole effects (Panel 2.11, Fig. 2.16b).
- From the second-order shift of the ^{23}Na and of the ^{87}Rb NMR lines the temperature behavior of the local order parameter has been deduced (Fig. 2.16c).
- The absolute estimate of φ from the measured value of ν_Q relies on the theoretical estimate of the electric field gradient and is thus somewhat uncertain but the relative temperature dependence is not.

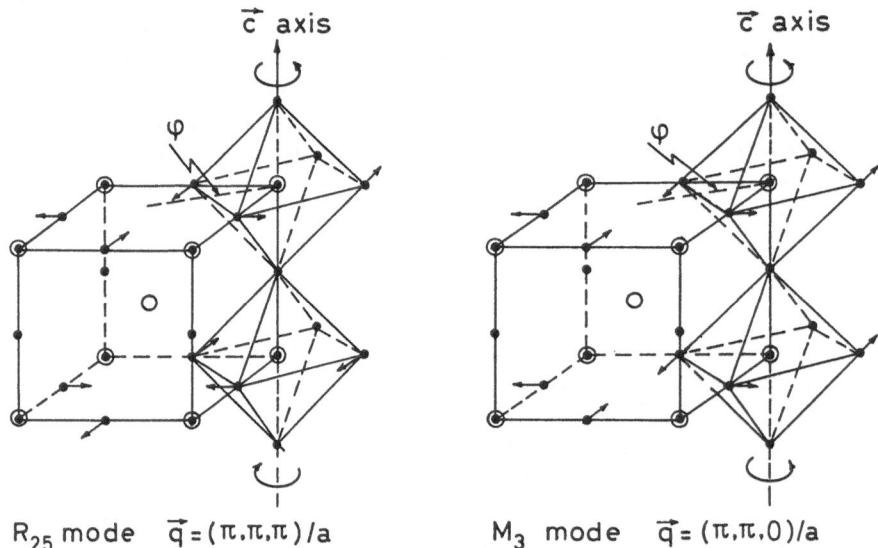

Fig.2.16a. Atomic displacements induced by the antiferrodistortive SPTs in perovskites

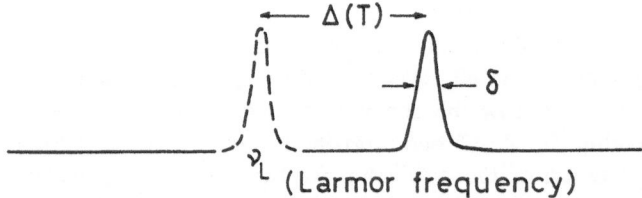

Fig. 2.16b. The second-order shift of the central NMR line is given by $\Delta(T) = A(I, \theta, \eta)\nu_Q^2/\nu_L$. For $I = \frac{3}{2}$ the maximum shift is $(10.8/16)\nu_Q^2/\nu_L$ for $\eta = 1$ and $(8.25/16)\nu_Q^2/\nu_L$ for $\eta = 0$ (axial symmetry)

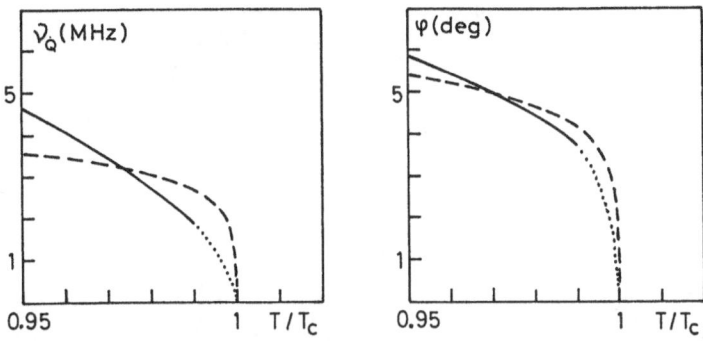

Fig. 2.16c. Quadrupole coupling constants and angle of rotation of the octahedra as a function of the reduced temperature: RbCaF$_3$(—), $\nu_Q \propto \sqrt{\Delta} \propto \varphi^2$ and $\varphi(T) \propto (1 - T/T_c)^{0.27}$; NaNbO$_3$(– –) $\nu_Q \propto \sqrt{\Delta} \propto \varphi$ and $\varphi(T) \propto (1 - T/T_c)^{0.17}$ (data from [2.41, 42])

PANEL 2.17 _____

NQR Study of the Order Parameter

- In K$_2$ReCl$_6$ the ferrodistortive SPT causes the rotation of the ReCl$_6$ octahedra (Fig. 2.17a).
- The Cl nuclei having $V_{zz} \| H_0$ give the ν^- and ν^+ resonance lines (Fig. 2.17b (i)); below T_c the line splits into two components (Fig. 2.17b (ii)); by rotating the crystal the two lines merge and $\varphi(T)$ is determined directly (Fig. 2.17b (iii)).

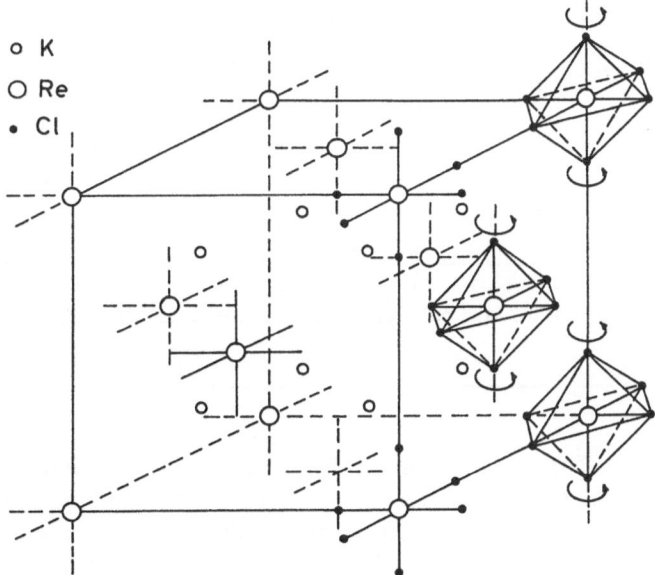

o K
O Re
• Cl

Fig. 2.17a. Schematic structure of K$_2$ReCl$_6$ showing the face-centered cubic arrangement of the ReCl$_6$ octahedra and the rotations occurring below T_c

117

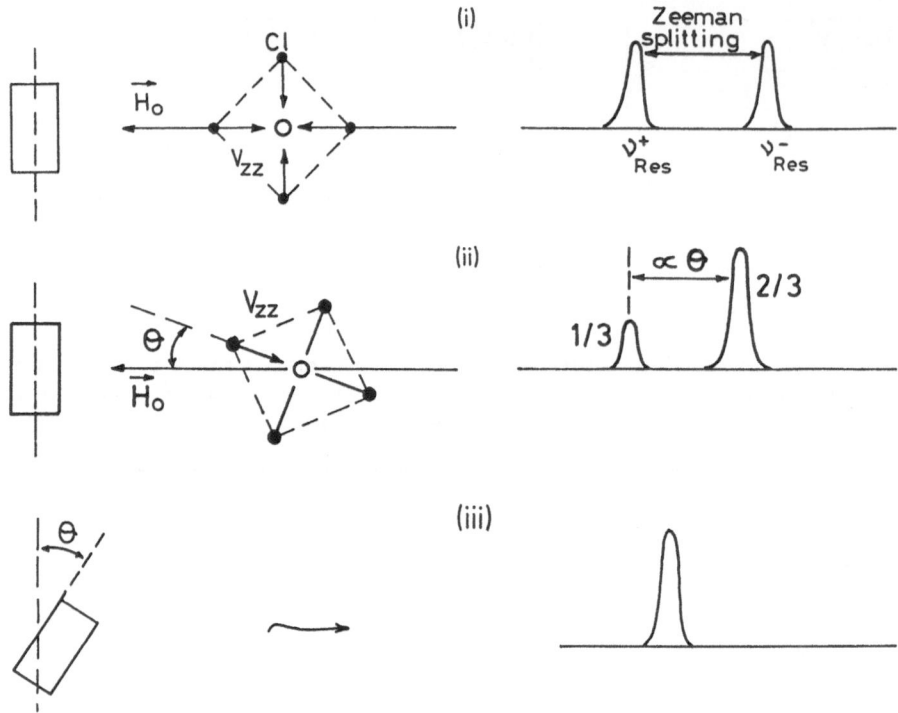

Fig. 2.17b. Cl NQR lines in K_2ReCl_6 and the direct determination of the local angle of rotation below T_c. The temperature dependence of the local order parameter is shown in Fig. 2.17c

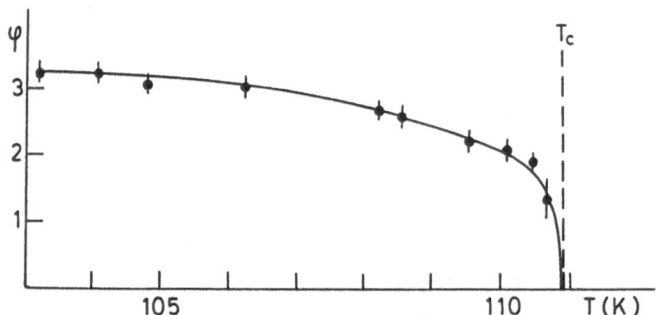

Fig. 2.17c. Temperature dependence of the local order parameter in K_2ReCl_6. ($\varphi = \theta$ in Fig. 2.17b.) From [2.43]

second-order SPT at 393 K. The order parameter is related to the probability of finding the N–C axis along the directions on a conical surface corresponding to the four equivalent positions of the C atoms in the ac plane, the positions becoming inequivalent below T_c. One of the frequency shifts turns out to be related to the square of the order parameter and a critical exponent $\beta = 0.250 \pm 0.005$ is inferred [2.44].

In a few cases the order parameter has been studied by chemical shift measurements. This is possible whenever the line separation between signals having different chemical shifts can be unambiguously related to the local order parameter. An example of this type is offered by the ^{13}C chemical shift in the layered squaric acid ($H_2C_4O_4$) in the vicinity of the ferroelectric type phase transition (Panel 2.18). The temperature behavior of certain line separations in the NMR spectrum shows a discontinuous jump at T_c accompanied by hysteresis leading to a claim of a first-order phase transition [2.45]. A nice feature of chemical shift measurements at structural phase transitions is that they can detect possible changes in the electronic structure at T_c.

PANEL 2.18 _____

Chemical Shift Study of the Order Parameter

- Squaric acid ($C_4H_2O_4$) is formed by planar C_4O_4-groups linked by hydrogen bonds to four neighboring C_4O_4-groups, thus yielding a molecular layered compound (Fig. 2.18a). While above T_c there are two ^{13}C NMR lines, in the low temperature phase one has four inequivalent carbon nuclei, depending on the four possible positions of the hydrogen in the bonds. Thus there are different chemical shifts, yielding four NMR lines (Fig. 2.18b). Since the line separations $\Delta\sigma_{24}$ and $\Delta\sigma_{31}$ are proportional to the order parameter (a measure of the net moment within the layer), its temperature dependence can be derived (Fig. 2.18c).

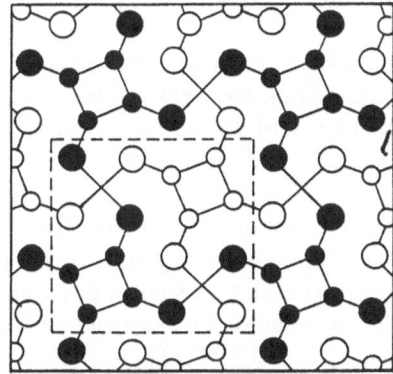

Fig. 2.18a. The planar structure of $C_4H_2O_4$

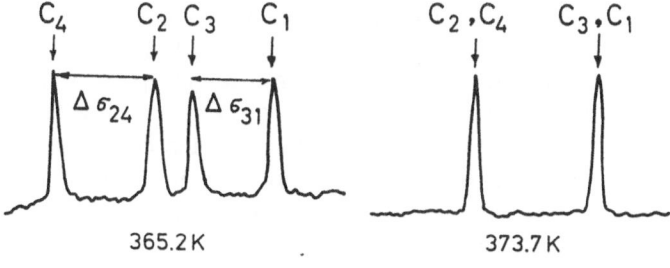

Fig. 2.18b. Examples of ^{13}C NMR spectra above and below T_c

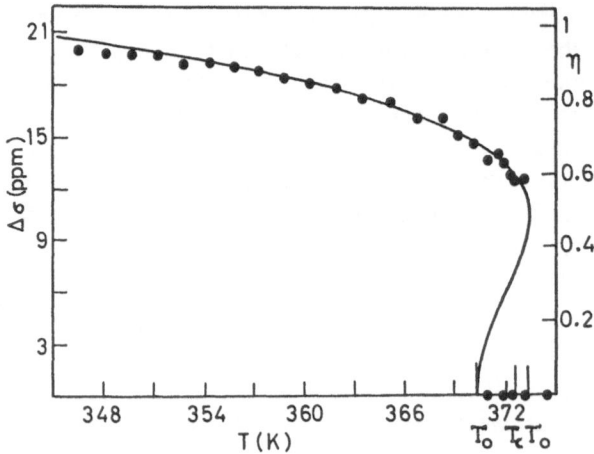

Fig. 2.18c. Temperature dependence of the order parameter in squaric acid as deduced from ^{13}C NMR chemical shift [2.45]

2.3.2 Symmetry and Anisotropy of Soft Modes

The critical dynamics accompanying the antiferrodistortive SPTs in perovskites is governed by the zone-boundary soft modes. The low temperature distorted phase results from the "freezing-in" of the soft mode. Near a SPT, even though the fluctuations may cross over from phonon-like excitations to order-disorder-type large-amplitude motions, (Panel 2.2), the symmetry and the anisotropy of the critical rotational fluctuations close to T_c are still related to the symmetry of the soft mode. This point is illustrated very nicely in the temperature behavior of the spin-lattice relaxation rate in a series of perovskites.

The relaxation rate near the transition can be related to the spectral densities of the critical dynamics by a semiclassical approach based on the weak collision theory (Panel 2.14). In determining the temperature dependence of T_1 and $T_{1\rho}$, an important role is played by a q-dependent coupling coefficient. By referring to the quadrupole interaction as the mechanism driving the relaxation process, one can relate the time-dependent EFG to the local fluctuation angle $\varphi(t)$ of the octahedra. In expressing the correlation function of the EFG components (as required by time-dependent perturbation theory in the weak collision approach) one has to sum both auto and pair correlation terms, over different lattice sites, thus obtaining a q-dependent factor A_q in the relationship of the relaxation rate to $S(q, \omega)$ (Panel 2.19). This circumstance is often advantageous because the integrated spectrum of the fluctuations over all wave vectors q becomes weighted by the A_q factors, which in turn are very sensitive to the symmetry and the anisotropy of the fluctuations. When the symmetry of the fluctuations of $\varphi(t)$ around the resonant nucleus is such that $A_{q_c} \neq 0$, the relaxation rate is dominated by the critical dynamics at $q = q_c$ and a divergent peak in T_1^{-1} is generally observed as a result of the critical slowing-down. On the other hand, when $A_{q_c} = 0$ the relaxation process is no longer dominated by the critical mode and in deriving the temperature and frequency dependence of T_1^{-1} the effects of the fluctuations at different wave vectors have to be considered in detail. A possible result could be the suppression of the divergence in T_1^{-1}.

120

Relaxation Rate Near a Phase Transition

- On approaching T_c the low frequency range of the power spectrum of the fluctuations increases for q close to the critical wavevector q_c (Fig. 2.19a).
- The relaxation rate can be written

$$T_1^{-1} \simeq \sum_q A_q S(q, \omega_L) \simeq A_{q_c} \sum_q S(q, \omega_L)$$

and, in accordance with the dynamical scaling,

$$S(q, \omega) = \langle |s_q|^2 \rangle \frac{2\pi}{\Gamma_q} f_1(\omega/\Gamma_q) \quad .$$

With the possible exclusion of a narrow temperature range close to T_c, the fluctuations are fast compared to ω_L and therefore $f_1(\omega_L/\Gamma_q) \approx 1$. Thus

$$T_1^{-1} \propto \sum_q \langle |s_q|^2 \rangle / \Gamma_q$$

and, as a consequence of the enhancement ($\langle |s_{q_c}|^2 \rangle \to \infty$) and of the slowing-down ($\Gamma_{q_c} \to 0$) of the critical fluctuations, one has a divergence of T_1^{-1} for $T \to T_c^+$ (Fig. 2.19b).

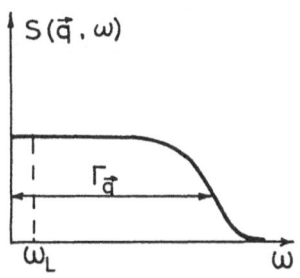 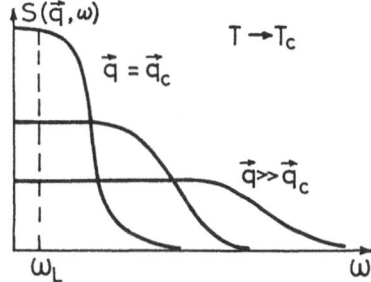

Fig. 2.19a. Frequency distribution of the dynamical structure factor for $q \approx q_c$ and its behavior on approaching the transition

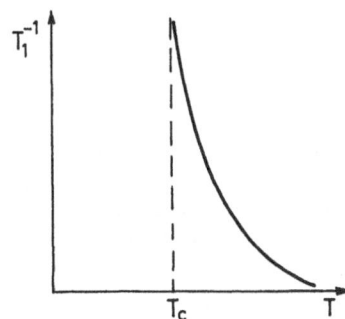

Fig. 2.19b. Sketch of the temperature behavior of the relaxation rate on approaching the critical temperature

Panel 2.20 offers a schematic representation of different cases that have been encountered in studying the antiferrodistortive transition in perovskites [2.38]. We would like to stress that we have purposely restricted here the illustrative analysis of the NMR results in perovskites to the role of the symmetry of the critical fluctuations. In fact the analysis of the critical temperature behavior in terms of the nature of the low frequency part of $S(q,\omega)$ involves the problematics of the central peak and its generation as a result of damping and non-linearity. Therefore we feel it more appropriate to defer that analysis to Sect. 2.6.

PANEL 2.20 _____

Effect of Symmetry and Anisotropy of Soft Modes on the Relaxation Rates

- The experimental findings for the spin-lattice relaxation rates of the A nucleus in ABX_3 perovskites reflect the symmetry of the soft modes. In fact the results sketched in Fig. 2.20 can be reproduced by the following symmetries:

	Symmetry	*Anisotropy*
$SrTiO_3$	R_{25}	$\Delta = 1$
$RbCaF_3$	R_{25}	$\Delta = 1/50$
$NaNbO_3$	M_3	$\Delta = 1/50$

where Δ is the anisotropy parameter in the generalized susceptibility $\chi(q,0) = \chi(q_c,0) \cdot \kappa^2 / [q^2 + \kappa^2 - (1-\Delta)q_z^2]$; $\Delta = 1$ corresponds to isotropy, and $\Delta \to 0$ to quasi-2D correlation in the critical rotational fluctuations (q is measured starting from q_c).

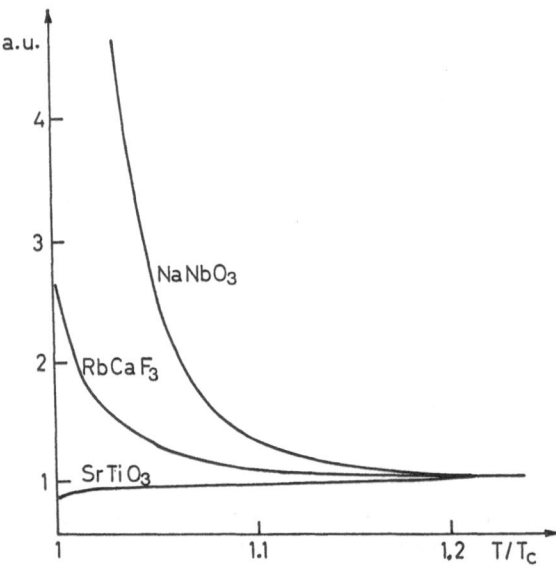

Fig. 2.20. Relaxation rates of the A nucleus (normalized to the high-temperature values) as a function of the reduced temperature, in the high-temperature phase of ABX_3 perovskites

2.3.3 Order-Disorder Critical Dynamics

In order-disorder phase transitions the essential feature is the presence in the power spectrum of the excitations of a low frequency part dominated by a soft diffusive mode. This low frequency part of the power spectrum can be investigated by T_1 and $T_{1\varrho}$ measurements in NMR and NQR experiments. As a guide for the analysis of the results it is convenient to specialize the dynamic structure factor $S(q, \omega)$ somewhat, either by referring to a microscopic model or by using the semi-phenomenological scaling approach [2.5] (Panel 2.19). As illustrative examples we discuss in the following some NMR results [2.46] in KDP-type crystals and in the molecular crystal of para-terphenyl [2.47]. A microscopic model suitable for describing uniaxial ferroelectric crystals is the dynamical Ising model, where the dynamics is characterized by the correlation time τ_{q_c} for the collective fluctuations at the critical wave vector q_c (in some cases a correction factor is introduced to take into account anisotropy effects) [2.38]. This model offers a good basis for discussion and interpretation of the data, also when tunneling is relevant, provided that a sizeable damping is present or that the resonant pseudo-spin waves, which describe the dynamics of the coupled tunneling states, have a central peak tail at low frequency. The formulas for $S(q, \omega)$ and for the nuclear spin lattice relaxation rates are summarized in Panels 2.2, 3, 14 and 15.

When a specific microscopic model is not at hand or one prefers to rely on a more general description including also the possibility of having a central peak, a scaling approach can be used [2.5].

a) KDP-type Crystals. The KH_2PO_4(KDP)-type crystals belong to a large class of the general formula MH_2RO_4 where M stands for K, Rb, Cs or NH_4 and R stands for P or As. For M = K, Rb or Cs the crystals are ferroelectric, while for $M = NH_4$ they are antiferroelectric. The ferroelectric properties are related to the position and the dynamics of the hydrogen bond in an O–H–O bridge whereby the phase transition is mostly order-disorder. Since the hydrogen atoms directly experience the critical dynamics they are ideal tools for NMR investigations. The deuteration of the crystals offers the additional advantage of allowing experiments on deuterum (D) nuclei ($I = 1$) which experience quadrupole interactions. Illuminating results have been obtained on the shape of the potential in which deuterons move, on the atomic displacements at the transitions and on the critical dynamics [2.48]. The increase of the correlation time on approaching T_c is directly reflected in the temperature behavior of the relaxation rate of deuterons, of P, and of the K (or Cs). The slowing-down of the collective motions increases the component of the spectral densities of the EFG functions at ω_L, thus causing a divergence of the relaxation rate driven by the fluctuating part of the quadrupole interaction. This is shown, for instance, in the divergence of the ^{133}Cs relaxation rate in CsH_2PO_4 near the order-disorder ferroelectric transition at 156 K (Panel 2.21). The relaxation rate has a temperature behavior which depends on the pseudo-one-dimensional nature of the fluctuations. Thus it was possible to estimate the temperature dependence of the proton intra-bond correlation time.

Relaxation Rate for Order-Disorder Phase Transitions

- In the light of the dynamical Ising model $S(q,\omega)$ can be written

$$S(q,\omega) = \frac{k_B T \chi(q,0) 2\tau_q}{1 + \omega^2 \tau_q^2}$$

  where $\chi(q,0)$ is the static generalized susceptibility and τ_q the collective correlation time.
- In CsH_2PO_4 the collective critical dynamics is of Ising type and can be considered one dimensional (Fig. 2.21a).
- The susceptibility can be evaluated exactly for the intra-chain dynamics, while the inter-chain effects can be treated in the mean field approximation. Thus, from the temperature behavior of the Cs relaxation rate, the proton intra-bond correlation time $\tau_{q=0}$ can be deduced (Fig. 2.21b).

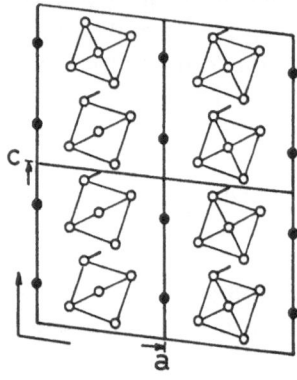

Fig. 2.21a. Schematic structure of CsH_2PO_4 evidencing how the collective dynamics of proton intrabond jumping is largely one-dimensionally correlated

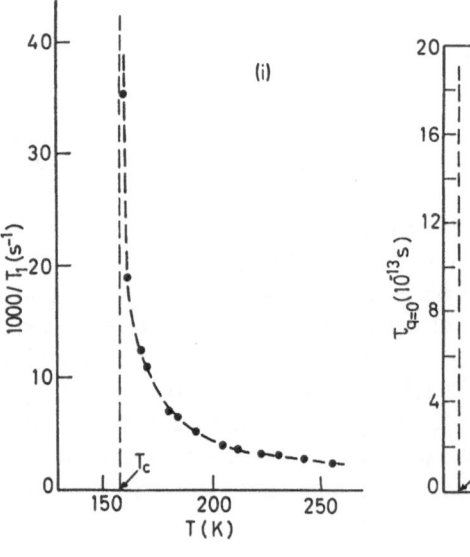

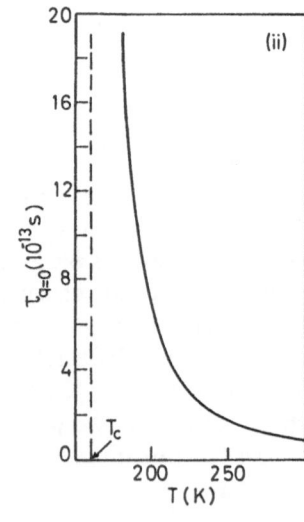

Fig. 2.21b. Experimental behavior of the Cs relaxation rate in CsH_2PO_4 (*i*) and the schematic divergence derived for the collective $q = 0$ correlation time (*ii*). Data from [2.46]

b) p-terphenyl. p-terphenyl ($C_{18}H_{14}$) is a molecular crystal belonging to the series of p-polyphenyls where interesting phase transitions occur. In p-terphenyl the phenylene ring is in a double-well potential resulting from the competition between the orthohydrogen repulsion and the intermolecular interactions (Panel 2.22). At $T_c = 190\,K$ an order-disorder antiferrodistortive transition twists the phenylene ring out of the plane formed by the two end phenyl groups. The critical dynamics of the order-disorder transition manifests itself in a pronounced divergence on approaching the transition according to a power law of the form $T_1^{-1} \propto (T - T_c)^{-n}$ with a critical exponent $n \sim 1$ (Panel 2.22). From the temperature behavior of T_1 and from the evaluation of the coupling coefficients associated with the nuclear dipolar interaction it was possible [2.47] to extract the temperature dependence of the decay rate Γ_c of the critical fluctuations (Panels 2.19 and 2.22 for the connection between T_1 and Γ_c). It should be noticed that the observed critical slowing-down of Γ_c reaches, near T_c, the order of magnitude of ω_L, thus leading to a slight frequency dependence of T_1. As shown in the panel there is good agreement between the results for Γ_c provided by T_1 data in the ordinary crystal and the few values obtained via the neutron back-scattering technique in the deuterated crystal.

2.4 Collective Behavior and "Phase Transitions" in the Presence of Positional Disorder

In this section we discuss the static properties and the related percolative-type problems involved in the behavior of the order parameter, as well as the local modes associated with disorder and/or strong anharmonicity in systems where the disorder is due to a random distribution of the units or to a local random modification of the interactions among units. Examples of NMR–NQR studies will be presented for systems ranging from mixed crystals retaining the main features of the ordinary phase transitions to strongly disordered systems where cooperative freezing occurs as a consequence of disordering dynamics. The first part will emphasize quasi-static phenomena while the second part deals with the dynamics of disordered units.

2.4.1 The Problem of the Order Parameter

Positional disorder and/or interaction disorder can be introduced by mixing isomorphous crystals A and B. When the two crystals both undergo transitions of the same symmetry with well-defined critical temperature T_c^A and T_c^B, the mixed system usually still undergoes a phase transition with a uniform long-range order parameter, as in the parent crystals [2.4]. This is the case, for example, with the isotopically disordered ferroelectrics HCl-DCl or KDP-DKDP as well as for the mixed order-disorder ferroelectric $Na(NO_2)$–$Ag(NO_2)$. The major interest is usually in the modifications of the critical dynamics and in the concentration dependence of $T_c^{(A+B)}$. An example [2.49] of an investigation of the concentration dependence of the transition temperature in mixed crystals studied by NQR is given in Panel 2.23 for the A_2BX_6 family undergoing a cubic to tetragonal transition. The behavior of $T_c = T_c(x)$ is

Critical Slowing-Down from Spin-Lattice Relaxation

- p-terphenyl, $C_{18}H_{14}$, is a molecular crystal with the phenyl rings in a double well potential. The order-disorder torsional oscillations bring the proton from A to B positions (Fig. 2.22a).
- In view of the two-dimensional character of the correlation, the characteristic frequency of the fluctuations Γ_c can be directly related to the proton T_1 (Panel 2.19):

$$T_1^{-1} \propto \sum_{\bar{q}} \langle |s_{\bar{q}}|^2 \rangle / \Gamma_{\bar{q}} \propto \Gamma_c \quad .$$

From the experimental data of T_1 vs T, Γ_c can be derived (Fig. 2.22b).

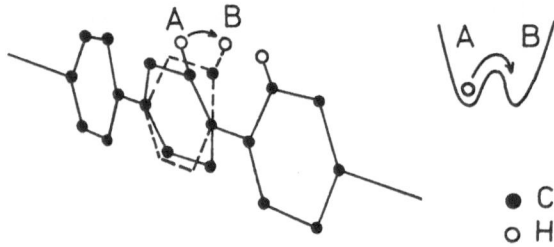

● C
○ H

Fig. 2.22a. The structure of a phenyl ring in p-terphenyl and the order-disorder motion of protons

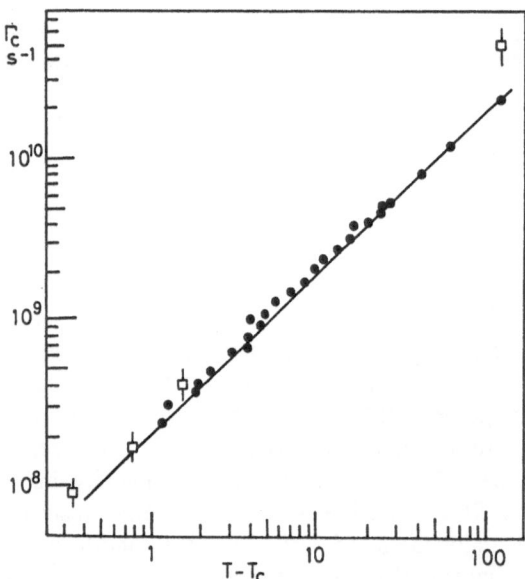

Fig. 2.22b. Critical behavior of the characteristic frequency of slowing-down in p-terphenyl as deduced from proton T_1 (●) and from high resolution neutron backscattering in the deuterated crystal (⌀). From [2.47]

NMR–NQR in Mixed Crystals with Uniform Long-Range Order

- The transition temperature $T_c(x)$ in mixed crystals A_2BX_6 (whose structure is shown in Fig. 2.17a) can be obtained from the changes in the Cl NQR frequencies (Fig. 2.23a). The experimental findings can be fitted in the framework of a VCA description (Fig. 2.23b).
- In $KMnF_3$, the replacement of K with Rb produces a decrease of the transition temperature from the cubic to the tetragonal phase. In Fig. 2.23c the evaluation of T_c from the NMR spectra is exemplified; in Fig. 2.23d the x-dependence of T_c is shown.

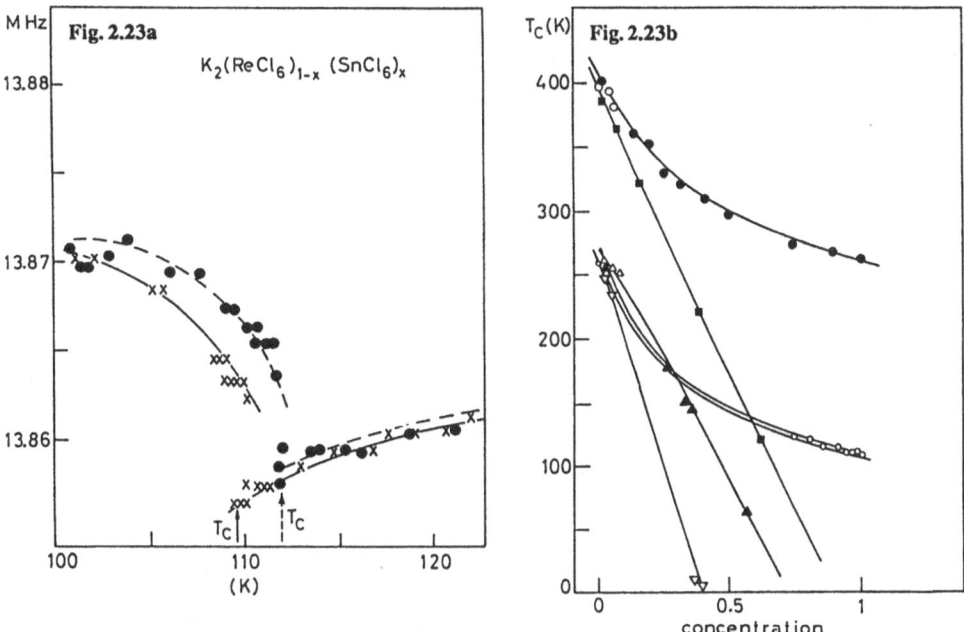

Fig. 2.23a. Temperature behavior of the ^{35}Cl NQR frequency for two crystals of the $K_2(ReCl_6)_{1-x}$ $(SnCl_6)_x$ family at $x = 0$ ($\times$) and $x = 0.08$ ($\bullet$). From [2.49]

Fig. 2.23b. Concentration dependence of the transition temperatures for various mixed crystals of the A_2BX_6 family. The data from NQR (*open symbols*) agree well with those obtained from DTA, DSC, and Raman scattering (*full symbols*). The solid lines are the theoretical best fits of $T_c(x)$ according to the VCA description [2.49]

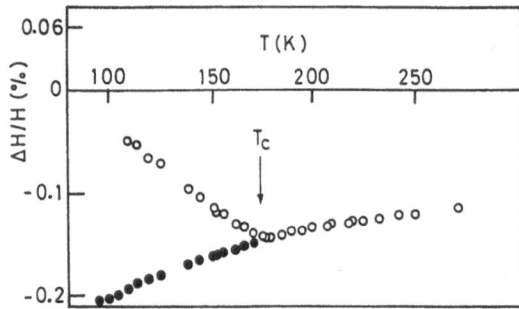

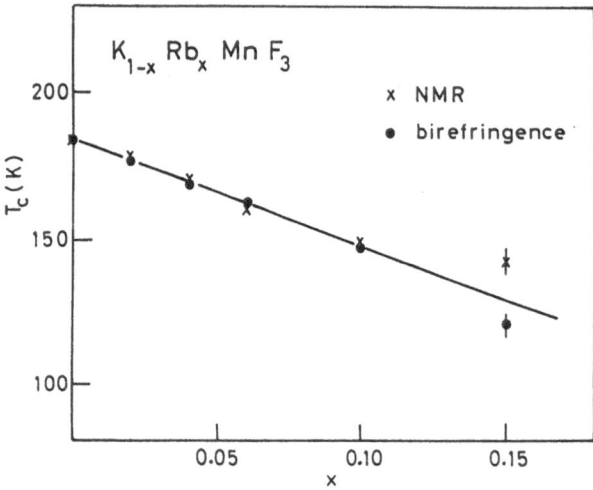

Fig. 2.23c.
Temperature behavior of the ^{39}K NMR resonance line at 3 MHz showing the splitting occurring at T_c for $K_{0.98}$ $Rb_{0.02}$ MnF_3 [2.50]. The two lines correspond to domains with $H_0 \| c$ (*solid circles*) and to domains with $H_0 \perp c$ (*open circles*) respectively and are associated with the angular dependent second-order quadrupole effects (Panel 2.11)

Fig. 2.23d. x-dependence of the transition temperature in $K_{1-x}Rb_xMnF_3$ as derived from ^{39}K NMR and birefringence measurements [2.50]

well described by the theoretical relation $T_c = T_c(0)(1+ax)(1+bx)^{-1}$, where a and b are fitting parameters based on a mean field description of random systems (virtual crystal approximation: VCA) [2.49].

When one of the two components, say B, does not exhibit a phase transition, then by adding B to A one may observe a progressive decrease of the transition temperature $T_c^{(A+B)}$. A good example of such an effect, studied through NMR measurements, are the RbMnF$_3$–KMnF$_3$ mixed crystals. KMnF$_3$ exhibits a structural phase transition driven by zone-boundary soft modes, while RbMnF$_3$ remains cubic at all temperatures. The second-order quadrupole shift of the ^{39}K central line was used [2.50] to monitor the onset of the tetragonal phase at $T_c(x)$ (Panel 2.23). For concentrations $x \lesssim 0.15$ of RbMnF$_3$ a phase transition to a long-range distorted phase occurs, but with a spatial modulation of the degree of tetragonality, as inferred from the broadening of the NMR line.

When the concentration of B reaches a certain critical range, the transition to the normal "ordered" phase ceases altogether. In this critical range novel effects, associated with a non-uniform order parameter, metastability and glass-type behavior can arise. The related phenomena are extremely complicated but NMR studies can nevertheless provide some illuminating information about the distribution and dynamics of random fields and order parameters, the resonance frequency of a given nucleus depending on the local field. Cases of this second kind are found in mixing the ferroelectric crystal $KNbO_3$ with paraelectric ($T_c \rightarrow 0$) $KTaO_3$ (Panel 2.24). Also mixed crystals $NaCN_x$–Cl_{1-x} and H_2–D_2 can be considered within this category. Of particular interest is the solid solution of ferroelectric RbD_2PO_4 and antiferroelectric $ND_4D_2PO_4$, which has been claimed [2.51] to represent frustrated H-bonded systems with competing ferroelectric and antiferroelectric interactions somewhat similar to magnetic spin-glasses. Here the distribution of random fields has been extracted from the difference between the inhomogeneous and homogeneous NMR line shapes [2.52].

a) $KNbO_3$–$KTaO_3$. The phase diagram of the mixed crystals KNb_xO_3–$KTa_{1-x}O_3$ indicates that for concentrations x of the ferroelectric active ion Nb down to about 5%, one still has the three ferroelectric phases of the prototype $KNbO_3$ (Panel 2.24). Below a concentration around the above value (and difficult to estimate more precisely) the complicated microscopic features of the low temperature phases have been satisfactorily clarified by recent NMR studies.

In Panel 2.24 the ^{93}Nb NMR spectra are shown schematically. In pure $KNbO_3$ and in the high Nb concentration crystal the onset of quadrupole effects at T_c is consistent with the occurrence of ferroelectric transitions with uniform long-range order in the Nb displacements (Sect. 2.2.1).

In the low Nb concentration sample (2% of Nb) the Nb NMR spectrum shows the unshifted central line and slightly inhomogeneously broadened satellite wings. The first-order quadrupole effects can be attributed to the slight asymmetry introduced by the presence around a given Nb of other Nb ions replacing the Ta's in neighboring cells. Below a given temperature, "T_c", the satellite wings disappear altogether suggesting a loss of local cubic symmetry at the Nb site. Correspondingly the intensity of the ^{181}Ta NMR central line signal drops abruptly to a lower value. This latter effect can be interpreted as a wipe-out effect around each off-center Nb ion. In fact, below T_c the signals from Ta nuclei within this volume become undetectable in view of the large shift inducded by the strong quadrupole effects. From the ratio of the signal intensities above and below T_c as a function of Nb concentration, the wipe-out volume was calculated to contain about 100 Ta ions [2.54]. One is led to the conclusion that the non-cubic symmetry is limited to the Nb bearing cells, without a uniform long-range cooperative distortion of the whole lattice (pseudo-ferroelectric phase) [2.54, 55].

b) $Na(CN)_x$–Cl_{1-x} and Isomorphic Compounds. These mixed crystals offer a good example of systems that, on diluting the elastic quadrupoles CN, cross over from an ordered array exhibiting orientational phase transitions (NaCN) to a random quadrupolar system with collective glass-like properties. The region of extreme dilution where the CN ions are almost totally decoupled has also been investigated by NMR.

In pure NaCN (or RbCN) one observes [2.56]: (i) the onset at T_c of static quadrupole effects consistent with tetragonal symmetry, arising from the freezing of

NMR in Mixed Crystals with Non-uniform Order-Parameter

- In the mixed crystals $K Ta_{1-x}Nb_xO_3$ the phase diagram (Fig. 2.24a) shows interesting features that have been clarified to a large extent by NMR (see below). In the region of high Nb concentration one observes three ferroelectric phase transitions with a uniform long-range order parameter, as in pure $KNbO_3$. For low Nb concentration one can define a "critical region" having the following properties: (i) non-uniform local order parameter; (ii) lack of true long-range order; (iii) glass-type effects. Data from [2.53].
- The ^{39}Nb NMR spectra display the behavior sketched in Fig. 2.24b,c and d [2.54].

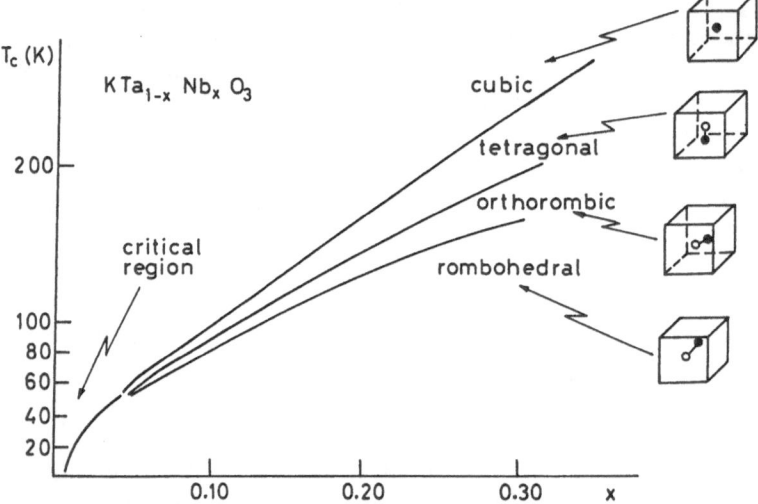

Fig. 2.24a. Phase diagram and schematic illustration of the off-center motions of Nb or Ta, in $KTa_{1-x}Nb_xO_3$

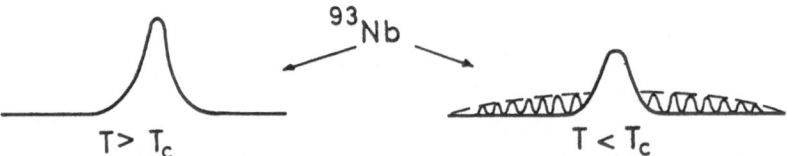

Fig. 2.24b. ^{93}Nb NMR lines in $KNbO_3$ above the transition (cubic phase, central and satellite lines superimposed) and below, where the first-order shift of the satellites resulting from the Nb collective off-center displacements causes a distribution of lines

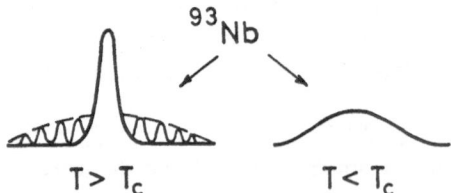

Fig. 2.24c.
^{93}Nb NMR spectra in $KTa_{0.64}Nb_{0.36}O_3$. In this high-concentration mixed crystal one has a first-order spreading of the satellites even in the cubic phase. At T_c the satellites disappear and the second-order shift, different for the three domains, gives a broadening of the lines

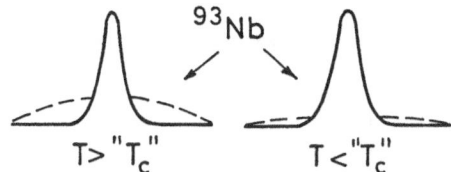

Fig. 2.24d. ^{93}Nb NMR spectra in KTa$_{0.92}$Nb$_{0.02}$O$_3$. In the low-concentration mixed crystals the ^{93}Nb satellite lines, which are only slightly broadened above the critical temperature, become unobservable below T_c

the CN dumbbells along a given pseudo-cubic direction, only head-to-tail reorientations remaining active; (ii) a small discontinuity in the spin-lattice relaxation rate at T_c; on cooling below T_c the temperature behavior of the relaxation rate indicates a monodispersive thermally activated reorientational motion of relaxational character, corresponding to head-to-tail reorientations.

Down to a concentration of the CN quadrupoles of $x \simeq 0.7$ the basic features of the phase transition remain unchanged. For $x \lesssim 0.7$ the first-order phase transition disappears. One observes glass-like cooperative effects whereby the CN ions experience a deformed local potential which leads to the formation of preferential directions for the orientation of the quadrupoles. This "ordering" phenomenon in a "disordered" system is marked by two specific NMR effects [2.57, 58]: (i) the ^{23}Na (and ^{35}Cl) spectrum is progressively broadened over a wide frequency range on decreasing the temperature; the amount of the broadening is a decreasing function of the CN concentration, with no detectable broadening for concentrations ($x \lesssim 0.2$) for which the quadrupoles are decoupled (Panel 2.25); (ii) the ^{23}Na spin-lattice relaxation rate changes its temperature behavior in a way related to the CN content, showing the presence of a wide distribution of correlation times and fast motion conditions [$\tau \lesssim \omega_L^{-1} = (2\pi \times 80\,\text{MHz})^{-1}$] down to liquid nitrogen temperatures (Panel 2.26).

PANEL 2.25

Distribution of EFG in Glass Phases

- In the mixed crystals (NaCl)$_{1-x}$(NaCN)$_x$ there is a glass region occurring for a concentration x of the CN dumbbells less than about 0.7 (Fig. 2.25a).
- The distribution of local EFGs at the Na or at the Cl sites resulting from a distribution of the preferential orientations of the quadrupoles and of their topological disorder causes the effects on the NMR spectra shown in Fig. 2.25b. The Gaussian distribution of the magnitude of the principal components heuristically assumed to interpret the broadening is

$$P(V_{aa}, V_{bb}) = A \exp\left(-\frac{V_{aa}^2 + V_{bb}^2 + cV_{aa}V_{bb}}{b^2}\right)$$

where $c = 1$ in order to preserve the condition of symmetry in a, b and c imposed by the average cubic symmetry [2.57, 58].

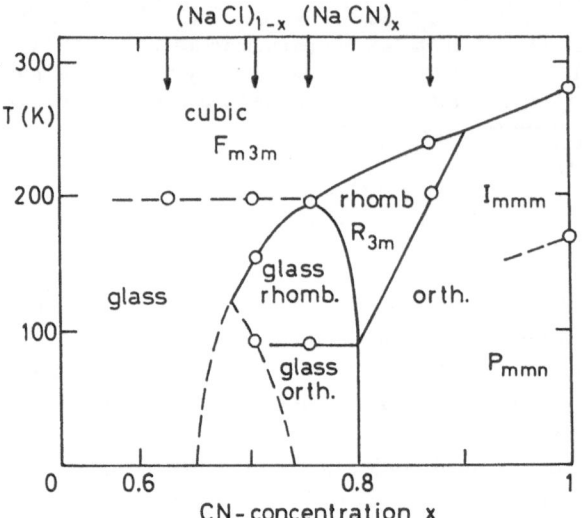

$(NaCl)_{1-x}$ $(NaCN)_x$

Fig. 2.25a. Phase diagram

cubic
F_{m3m}

rhomb
R_{3m}

I_{mmm}

glass

glass
rhomb.

orth.

glass
orth.

P_{mmn}

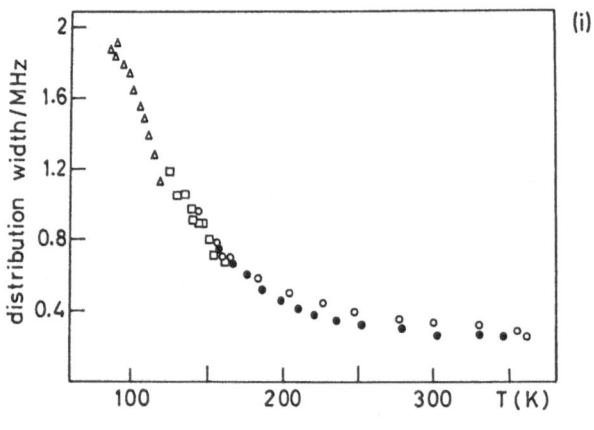

(i)

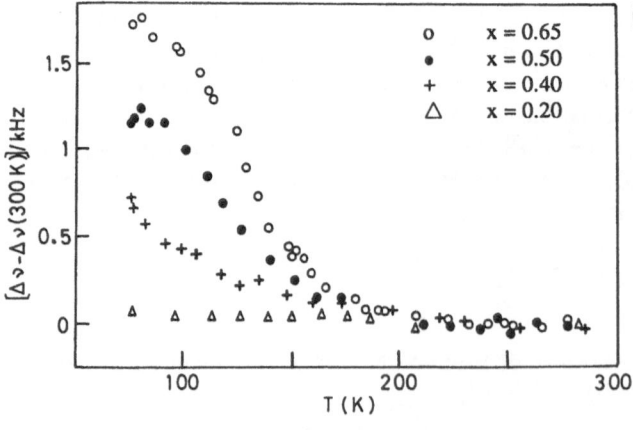

(ii)

o	x = 0.65
•	x = 0.50
+	x = 0.40
△	x = 0.20

Fig. 2.25b. Effects on the ^{35}Cl spectra in $Na(CN)_x Cl_{1-x}$ resulting from the distribution of local EFGs. The results for the distribution width (in MHz) (*i*) have been obtained for $x = 0.65$ from the central line and from the satellite line distribution, at various resonance frequencies. In (*ii*) one sees the effect on the distribution for various x [2.57, 58]

Distribution of Correlation Times in Glass Phases

- In pure NaCN, above and below T_c one finds approximately the behavior of the ^{23}Na relaxation rate characteristic of a monodispersive process, namely

$$T_1^{-1} \propto \frac{\tau_c}{\left(1 + \omega_L^2 \tau_c^2\right)}$$

with a minimum in T_1 for $\tau_c \sim \omega_L^{-1}$ (Panel 2.15).

In the presence of a distribution of correalation times one can write

$$T_1^{-1} \sim \int p(\tau) \left[\frac{\tau}{\left(1 + \omega_L^2 \tau^2\right)} \right] d\tau \quad .$$

- From the comparison of the temperature behavior of ^{23}Na T_1^{-1} in $Na(CN)_x Cl_{1-x}$, with the one characteristic of a monodispersive process (Fig. 2.26) the occurrence of a distribution of order parameters, typical of glassy phases, has been inferred.

  For a given temperature the average correlation time is shorter for a smaller CN concentration, while there is a continuous increase of both the distribution of the EFGs and of the effective correlation times.
- At low temperature the CN quadrupoles are practically frozen (quadrupole order) while the 180° head-to-tail reorientations are still present (dipolar disorder).

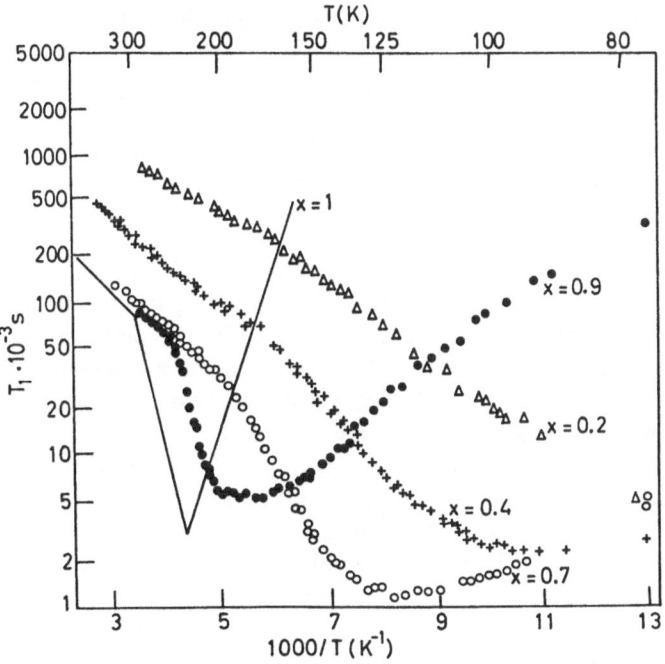

Fig. 2.26. ^{23}Na relaxation times vs $1/T$ in $Na(CN)_x Cl_{1-x}$ for various x, compared to the case for $x = 1$ [2.56, 57]

The interpretation of the broadening of the NMR spectra is given by relating the distribution of the EFGs in the Σ^P frame of reference (which is assumed to coincide with the crystalline cubic axes) to the preferential orientation of the CN quadrupoles along the [100] directions. By assuming a simple Gaussian distribution for the magnitude of the V_{JK} in Σ^P one can reproduce (after proper folding with the dipolar broadening) both the satellites and central line shape at a given temperature. Thus the temperature behavior of the width of the distribution is completely determined.

The analysis of the relaxation rate and temperature behavior starts by considering that in pure NaCN above $T_c \simeq 288\,\mathrm{K}$ the relaxation mechanism is due to the fast reorientations of the CN axis among equivalent cubic directions ([100] axes or [111] axes), which are thermally activated. Below T_c the dumbbells are frozen along a given direction and only head-to-tail reorientations occur. These motions give rise to the classical V-shaped behavior for T_1 vs T^{-1}, with a single thermally activated correlation time τ (Fig. 2.26).

In the concentration range corresponding to the glassy state the relaxation is driven at all temperatures by fast reorientations among the cubic directions. The random deformation in the local potential, related to the disorder, causes a non-cubic averaging of the reorientations, yielding an ellipsoidal deformation of the tensor describing the state of motion of the rotators. The degree of deformation from the spherical averaging can be related to an effective "order parameter" (Sect. 2.4.1c). The temperature and concentration dependence of the distribution of EFGs of at the Na^{23} sites reflects the corresponding dependence of the order parameter. Correspondingly, one has a distribution of the effective correlation times describing the fluctuating EFGs. The principal axes of the ellipsoids giving the directions of preferential orientation are bound to be oriented at random only along the directions of cubic symmetry of the underlying ionic lattice. Thus the glassy character involves mostly the values of the local "order parameter" which measure the extent of the preferential orientation.

We note that, in this respect, the glassy character is not fully equivalent to magnetic spin glasses where the decoupling of the spins from the lattice allows all possible random orientations for the average spin directions.

c) N_2 and Ortho-para H_2 and D_2. The molecular crystals of solid nitrogen and hydrogen offer ideal model systems for studying orientational ordering transitions in the presence of disorder. The disorder in these cases is induced by the presence of inert $J = 0$ para-rotators which dilute the effective $J = 1$ ortho-rotators. The strong interactions of isotropic character lead to the lattice structure, while the weaker electric quadrupole interactions H_{QQ} among the molecules determine the orientations of the molecular axes. NMR has given nice examples of the study of orientational ordering as a function of the concentration of ortho-para mixtures.

For a layer of N_2 physisorbed on graphite, the interaction energy H_{QQ} is of the order of the rotational energy $BJ(J+1)$, yielding a quasi-classical system with poorly defined rotational states. Thus the local order parameter $p_i(\gamma)$, defining the degree of preferential orientation in the motion of the molecular axis, involves the time average $\langle 3Z_i^2 - 1 \rangle / 2$ which describes the mean alignment along Z_i (Panel 2.27). The effective dipole-dipole Hamiltonian H_d (Panel 2.28), which is a perturbation of the Zeeman

Arrays of Quantum Rotators

- In H_2 or $^{15}N_2$ there are two molecular species with the characteristics:

$I = 1$	(ortho)	$I = 0$	(para)
χ	spin symmetric	χ	spin antisymmetric
$J = 1$	rotational state	$J = 0$	rotational state

- When the electric quadrupole interaction $H_{QQ} \approx (6e^2Q^2/25R_0^5) \ll BJ(J+1)$, then J is a good quantum number. In H_2, $H_{QQ} \approx 0.5\,K$ and $B = 60\,cm^{-1}$.
- When $H_{QQ} \sim B$ there are no well-defined rotational states and the system is quasi-classical (the case of N_2).
- The order parameters for the orientational phase transitions are given by the average of the spherical harmonics Y_{2m}:
  - i) thermodynamical averages for pure rotational states (H_2);
  - ii) time averages for quasi-classical systems (N_2).

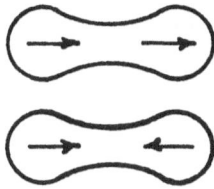

Fig. 2.27. Sketch of ortho (*upper*) and para (*lower*) molecules

Orientational Phase Transition in N_2 Layers

- In the registered phase of a layer physisorbed on graphite, the N_2 quadrupoles occupy the sites of a triangular lattice (Fig. 2.28a). The effective nuclear dipole-dipole $^{15}N - ^{15}N$ interaction, which is a perturbation of the Zeeman Hamiltonian, can be written

$$H_d = -\frac{\hbar D}{3}P_2(\theta_i)\left[3I_{z_i} - 2\right]p_i$$

where $D \simeq 8710\,rad\,s^{-1}$. In Fig. 2.28b ξ_i is the axis of preferential orientation and $p_i = \langle P_2(\gamma_i)\rangle_t$.

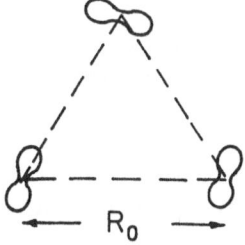

Fig. 2.28a. N_2 quadrupoles on a triangular lattice

- The echo amplitude following the $(\pi/2 - \tau - \pi)$ pulse sequence can be related to the order parameter $p(T)$ allowing one to trace its temperature behavior (Fig. 2.28c).

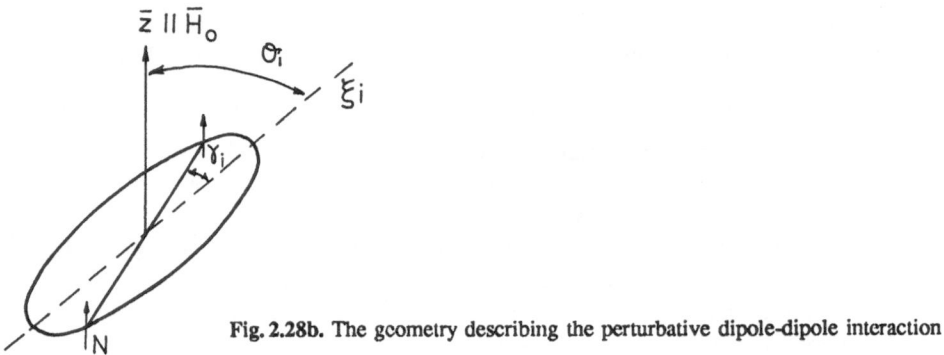

Fig. 2.28b. The geometry describing the perturbative dipole-dipole interaction

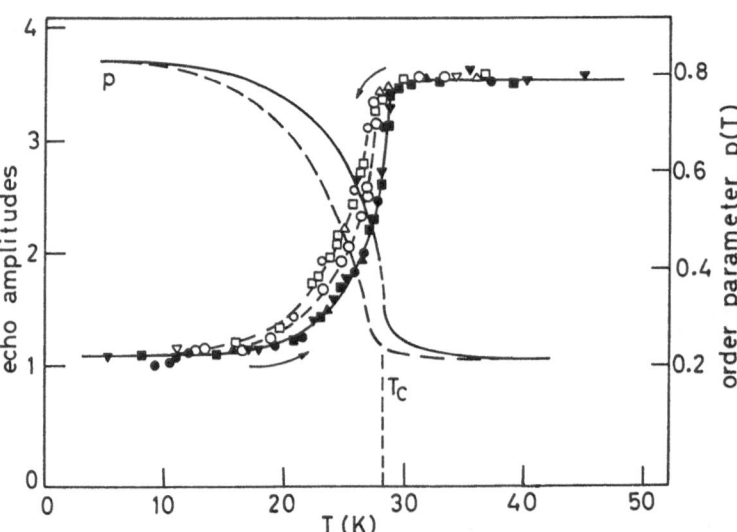

Fig. 2.28c. Echo amplitude and order parameter for N_2 on graphite; the slightly first-order character of the transition (possibly induced by the cubic symmetry of the fluctuations) is indicated by the thermal hysteresis and by the slight discontinuity. From [2.59]

Hamiltonian, can be written in terms of $p_i(\gamma)$. Thus the amplitude A of the ^{15}N NMR echo signal is a function of the order parameter. From the temperature dependence of A one derives the behavior of the order parameter around the transition ($T_c \simeq 28\,K$) to the ordered phase. In particular, the behavior of p was found to show thermal hysteresis and a small discontinuity [2.59]. These effects were used to support the hypothesis of a second-order phase transition (within MFA) driven slightly first-order

136

by the fluctuations, as expected for a system belonging to the same universality class as the $n = 3$ component Heisenberg model with face-oriented anisotropy.

In solid ortho-para H_2 and ortho-para D_2 mixtures one may reduce the concentration of the $J = 1$ quantum rotators below the value $x = 0.75$ characterizing pure H_2 (Panel 2.29). When x reaches a critical value $x_c = 0.55$, the long range orientational order is lost and the frustration associated with the combined effect of disorder and the anisotropy of H_{QQ} causes a random freezing. In this state of orientational disorder both the local order parameter $\sigma_i = \langle 3J_z^2 - 2 \rangle_T$ and the local symmetry axis z_i vary from site to site.

The dipole-dipole interaction is directly related to the order parameter, thus yielding a fine structure in the ^{1}H NMR spectrum. In fact, for $T > T_g$, where $\langle \sigma_i \rangle = 0$ one has a single narrow line; for an ordered state (for $x > x_c$) where $\langle \sigma_i \rangle = -2$ at each site the quadrupolarization is reflected in a spectrum with a Pake-type doublet; finally for the glassy state ($x \lesssim x_c$) the distribution of the local order parameters and of Z_i causes a broad NMR spectrum whose second moment is directly related to the glass order parameter equivalent to q_{EA} in spin glasses. This order parameter shows an increase on cooling below 400 mK, corresponding to a progressive increase in the glass quadrupolarization [2.60]. It is not clear if the change of slope of M_2 vs T around 100 mK marks a symmetry breaking characteristic of a true phase transition in a thermodynamical sense or not.

2.4.2 The Problem of Critical Dynamics

We turn now to the modification of soft mode behavior and to local dynamics in disordered systems. NMR has given unique information on critical dynamics, particularly because the dynamics of the interacting disorder units have a pronounced local character.

a) $KNbO_3$–$KTaO_3$. In pure $KNbO_3$ the ^{93}Nb relaxation is dominated by the ferroelectric soft mode which, on approaching T_c from above, can be viewed as a correlated motion of the Nb ions among the eight equivalent off-center positions along the main cubic diagonals [2.37]. The fact that the Nb motions are $q = 0$ correlated leads to a decrease of the inter-cell contribution to the fluctuating EFGs. Thus the relaxation rate decreases on approaching the cubic to tetragonal phase transition. In the low-Nb concentration mixed crystal $KTaO_3$–$KNbO_3$ (KTN), the Nb ions are well separated and the inter-cell contribution may be neglected. The intra-cell contribution, quadratic in the Nb displacement, is much less effective in driving the spin-lattice relaxation, as can be seen by comparing the order of magnitude of ^{93}Nb relaxation rates in pure $KNbO_3$ with those in $KTaO_3$:Nb 2.8 % [2.54] (Panel 2.30). The ^{93}Nb relaxation rate in KTN 36 % has a behavior similar to that in pure $KNbO_3$ [2.54].

The small anomaly observed at "T_c" [2.54, 55] can be ascribed to a "slowing-down" of the Nb motion, precursor effects of the transition to the pseudo-ferroelectric phase.

For the ^{181}Ta relaxation rate one can observe that the order of magnitude is the same as in $KTaO_3$:$KNbO_3$ at 2 % and in $KTaO_3$ and in both cases depends

Order Parameter in Disordered Quantum Rotators

- In the array of $J = 1$ quantum rotators H_2–D_2 mixture, for a concentration $x \lesssim 0.55$ of ortho H_2 molecules one has a glass-type phase (Fig. 2.29a).
- The local order parameter for the quantum rotator is

$$\sigma = \langle 3J_z^2 - 2 \rangle_T$$

the eccentricity $\eta = \langle J_x^2 - J_y^2 \rangle$ (Fig. 2.29b) usually being small.
- The second moment M_2 of the inhomogeneously broadened NMR line (Panel 2.10) (for $x < x_c$) is directly related to the order parameter

$$q = \left(\langle \sigma_i \rangle_{T,J} \right)^{1/2}$$

from the equation $M_2 = 9d^2 q^2 / 5$, while the dipole Hamiltonian is $\mathcal{H}_{dip} = d(3I_z^2 - 2)\langle 3J_z^2 - 2 \rangle_T$. The temperature behavior of q^2 is shown in Fig. 2.29c.

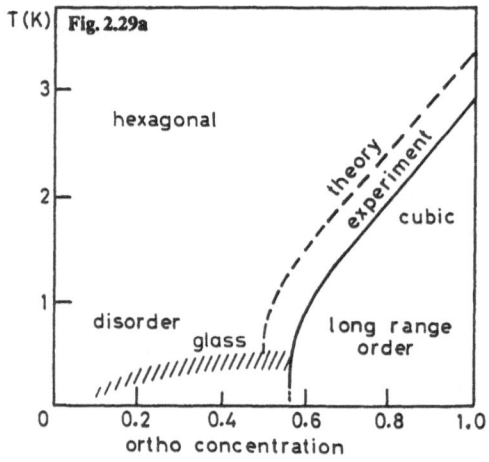

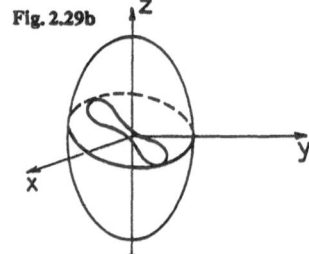

Fig. 2.29b. Sketch of a quantum rotator for the definition of the order parameter and of the eccentricity

◄ Fig. 2.29a. Phase diagram for the H_2–D_2 mixture evidencing the glassy phase

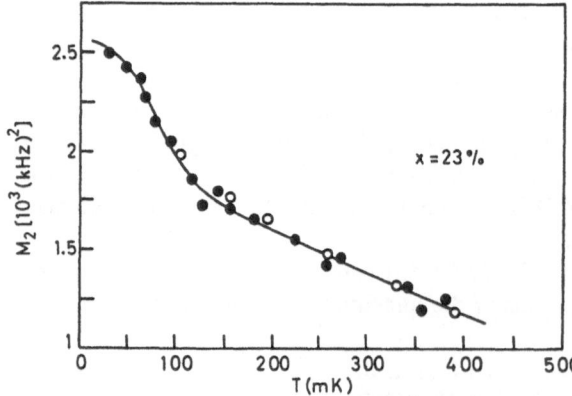

Fig. 2.29c. Second moment of the NMR line vs temperature in the glassy phase of H_2–D_2 crystals. From [2.60]

quadratically on T as expected for a Raman two-phonon process (Fig. 2.30b). Since the ^{181}Ta NMR spectra clearly indicate that below "T_c" some Ta ions still sit in a local cubic symmetry [2.55], one is led to conclude that the anomaly in the ^{181}Ta relaxation rate (as well as for ^{93}Nb) is associated only with the dynamics of Nb ions. This confirms the incapability of the "critical" dynamics of the Nb to drive a uniform long-range ferroelectric distortion in the host lattice.

PANEL 2.30 _____

Relaxation Studies of Soft Modes in Ordered vis-a-vis Disordered Crystals

- In pure KNbO$_3$ the ferroelectric slowing-down on approaching the cubic to tetragonal transition ($T_c \simeq 700$ K) produces a kind of antidivergence in the ^{93}Nb relaxation rate because of the effect of the symmetry of the soft mode on the A_q factors (Panel 2.19, Fig. 2.30a).
- In KTN mixed crystals, for low concentration of Nb (see the "critical region" in the phase diagram in Panel 2.24) the antidivergence associated with the $q = 0$ soft mode is replaced by an enhancement of T_1^{-1}, as indicated by ^{93}Nb and ^{181}Ta relaxation rates (Fig. 2.30b). The anomaly at "T_c" is related to the critical "slowing-down" of the Nb off-center motion. The Nb motion is the dominant relaxation mechanism for ^{93}Nb while it "sits" on the top of the Raman two-phonon mechanism for ^{181}Ta.

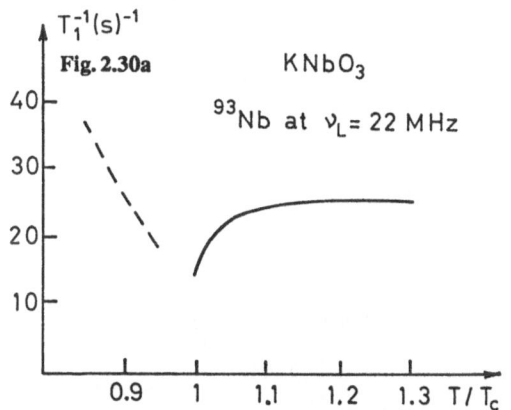

Fig. 2.30a. Sketch of the behavior of the Nb relaxation rate around the cubic-tetragonal SPT in KNbO$_3$ as resulting from the enhancement and slowing-down of the $q = 0$ soft mode. From [2.53]

Fig. 2.30b. ^{93}Nb and ^{181}Ta relaxation times vs T showing the effects due to the slowing-down of the Nb off-center motion

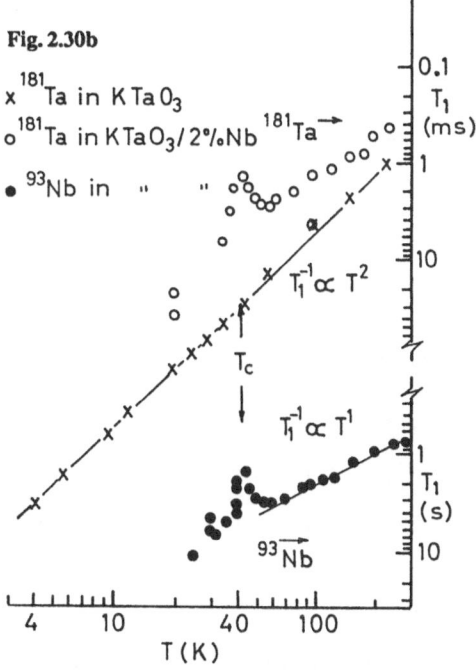

NMR Study of Local Modes in KTaO₃:Li

- The six equivalent off-center positions for the Li ion impurity are displaced from the center of the cubic cell by 1.26 Å. The height of the local energy barrier is about 1000 K (Fig. 2.31a).
- The slowing-down of the Li hopping correlation time causes the appearance of the first-order quadrupole-perturbed spectrum at a temperature where $\tau_Q \sim \nu_Q^{-1}$ (Fig. 2.31b). The simulation of the spectrum, obtained by using the correlation time given by dielectric measurements or by T_1 (Fig. 2.31c) agrees with the experimental findings.
- In the light of dielectric measurements one can describe the hopping motions of the Li impurities by a simple monodispersive relaxation process.

    One then has

$$T_1^{-1} \sim \langle |\nu_Q|^2 \rangle \frac{\tau_c}{1 + \omega_L^2 \tau_c^2}$$

with $\tau_c = 2 \times 10^{-14} \exp(1000/T)$ s.

The fit to the ^{7}Li, ^{39}K, and ^{181}Ta relaxation rates on the basis of this model is very good (Fig. 2.31c). The average mean square interaction $\langle |\nu_Q|^2 \rangle$, apart from the dependence on the properties of the nucleus, is the same for all concentrations x in the case of Li relaxation, as expected. For ^{39}K relaxation, however, where the relaxation mechanism is the inter-cell one, the relaxation rates depend on the concentration x through $\langle |\nu_Q|^2 \rangle$ (Fig. 2.31c (i)).

Fig. 2.31a

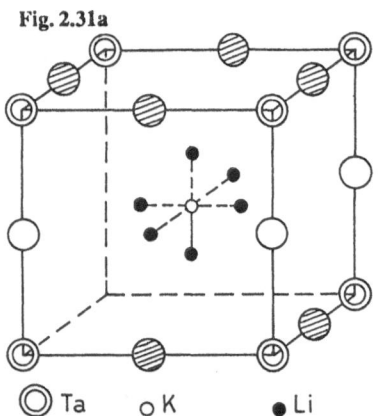

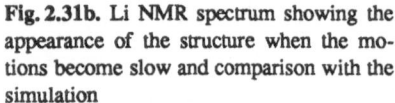

◎ Ta   ○ K   ● Li

Fig. 2.31a. The cubic unit cell around K in KTaO₃ and the out-of center position that can be taken by a Li substitutional impurity

Fig. 2.31b. Li NMR spectrum showing the appearance of the structure when the motions become slow and comparison with the simulation

Fig. 2.31b

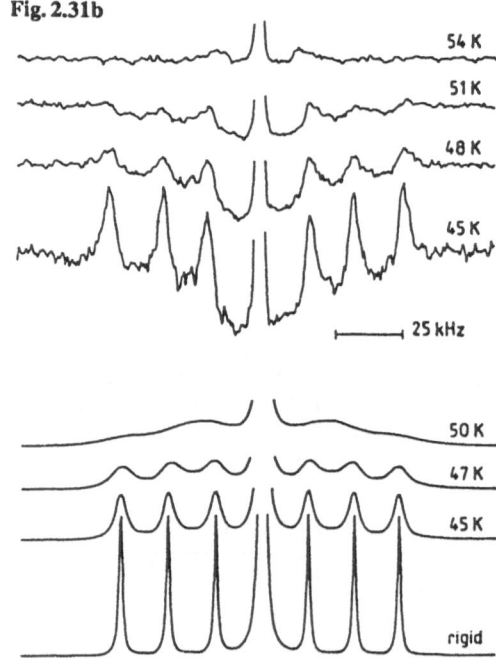

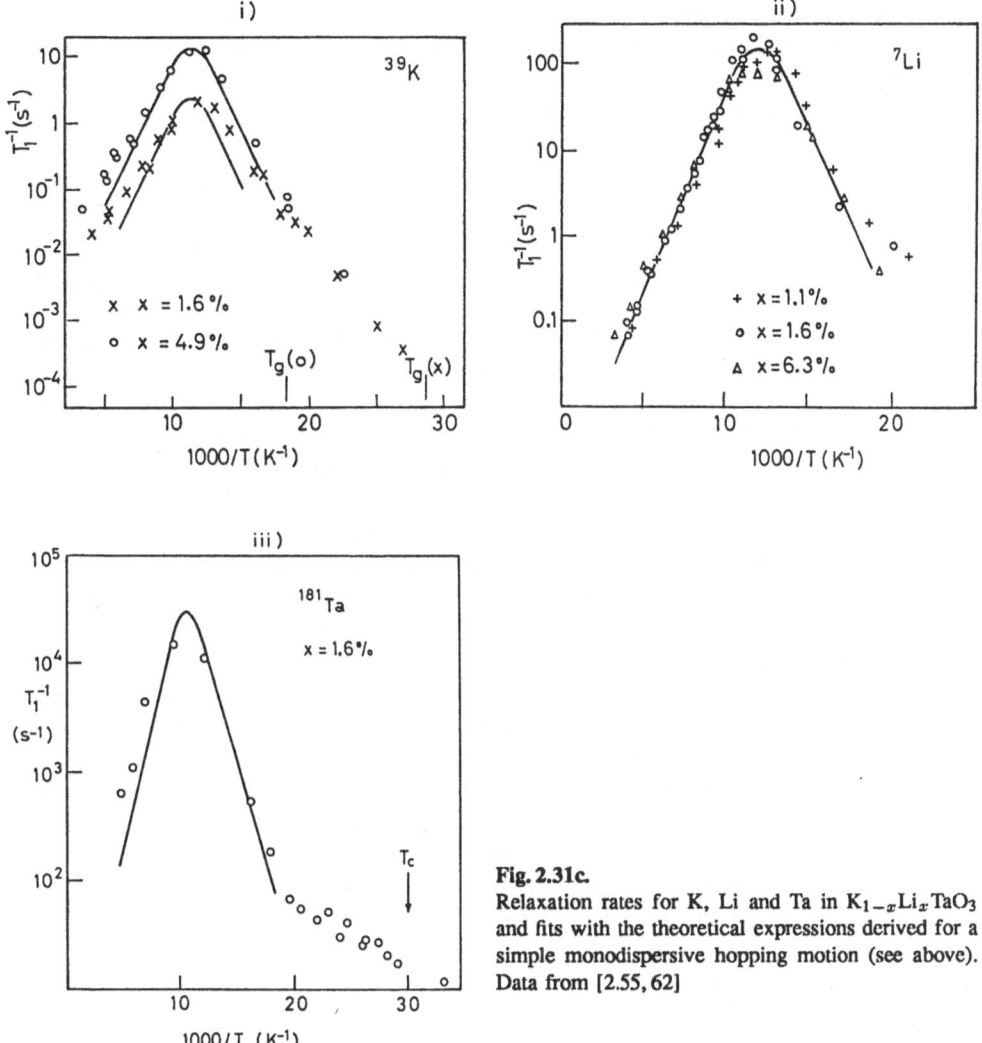

Fig. 2.31c.
Relaxation rates for K, Li and Ta in $K_{1-x}Li_xTaO_3$ and fits with the theoretical expressions derived for a simple monodispersive hopping motion (see above). Data from [2.55, 62]

b) KTaO₃:Li. The replacement of K by Li gives rise to an off-center ion with a local potential with a six-fold minimum. NMR and dielectric measurements have provided the values for the off-center displacement and for the energy barrier [2.61] (Panel 2.31).

In view of the occurrence of permanent dipole moments associated with the randomly distributed Li ions and the related frustration, the system represents potentially the electric dipolar counterpart of a magnetic spin-glass. For this reason

141

KTaO$_3$:Li has been extensively investigated and ^{7}Li, ^{39}K, ^{181}Ta NMR [2.55, 62] have contributed decisively to our understanding of static and dynamical properties. An important conclusion drawn from NMR measurements is that the Li ions are characterized by a thermally activated local hopping dynamics down to a time scale of about 10^{-4} s and below. In fact the ^{7}Li spectrum displays a single line, with no static quadrupole effects (thus average cubic symmetry) down to about 50 K. Below this temperature (only slightly dependent on the Li concentration) one observes the expected first-order quadrupole perturbed spectrum, not as consequence of a phase transition (as in ordered crystals) but simply as a consequence of the slowing-down of the hopping motion below the value $(70 \text{ KHz})^{-1} \simeq 10^{-5}$ s.

Furthermore, the behavior of the relaxation rate as a function of temperature indicates that the hopping motion around an average cubic position persists at much lower temperatures. It should also be stressed that the T_1 vs T data can be fitted without adjustable parameters using for the root-mean-square fluctuating interaction the static ν_Q and for the correlation times the ones obtained from dielectric dispersion at low Li concentration (Panel 2.31).

In order to decide about the problem of a possible cooperative formation of a dipolar glass-like state, analogous to a spin glass, one should observe either an equivalent of the Edwards-Anderson order parameter via the spectra or, at least, a cooperative dynamics which causes an anomaly in the temperature dependence of the local correlation times and thus on T_1. Neither of these effects are actually observed, leading to the conclusion that the Li local dynamics remains, up to a concentration of about 6 %, largely a single-particle dynamics weakly coupled to the host lattice (at a least for $T \gtrsim 35$ K).

On the other hand, the host lattice must undergo a structural distortion at a well-defined transition temperature, if one wants to explain the birefringence and Raman data [2.63], but this distortion is not coupled to the Li dynamics.

At very low temperatures, glass-like polar configurations associated with very slow cooperative dynamics of the Li ions is possible, as suggested by dielectric measurements [2.64]. In view of the very slow frequencies involved, this effect could be investigated by NMR techniques only by looking at relaxation processes in the rotating or dipolar frame ($T_{1\varrho}$, T_{1D}).

c) **KTaO$_3$:Na.** The substitution of Na for K in KTaO$_3$ leads to a situation rather different from the case of KTaO$_3$:Li. In fact, the NMR spectrum [2.65] indicates that the static quadrupole effects at the Na site are negligible at all temperatures; thus no permanent dipoles associated with the Na off-center positions can be invoked. One can consider the KTaO$_3$:Na system as a mixture of NaTaO$_3$ and KTaO$_3$, where pure NaTaO$_3$ has structural transitions starting at high temperatures (~ 1000 K).

One expects that, below a certain "critical" concentration of Na, estimated to be around 20 %, the conventional phase transition is inhibited and one may have, at low temperatures, a non-uniform local distortion, without a true long-range order. This situation is analogous to that found in KTaO$_3$:KNbO$_3$, for very low Nb concentrations.

In this context the Na ion sits in a strongly anharmonic potential and exhibits a complex dynamical behavior with several features characteristic of disordered systems (distribution of correlation times, metastability and resonant tunneling with a distribution of barrier heights).

At high temperatures the ^{23}Na relaxation rate (for $x < 0.20$) follows a T^2 law, is frequency independent, and is of the same order of magnitude as in NaCl [2.65, 66]. Below a temperature of about 150 K one observes a significant departure from the behavior expected for harmonic phonon modes. The deviation can be explained by assuming the onset of a strong, direct relaxation mechanism induced by a diffusive low-frequency component in the spectral density $J(\omega)$ of the motion, having a width $\tau^{-1} \sim \omega_L$ (Panel 2.32). This relaxational component in $J(\omega)$ can be obtained theoretically by considering the motion of interacting particles each experiencing a strongly anharmonic local potential (Panel 2.2). With regard to the Na spin-lattice relaxation, the long-time correlation function for the motion is equivalent to a hopping among the off-center minima in the anharmonic potential. This result is obtained even though, contrary to the case of Li, the mean square displacement of Na is larger than the separation between the minima.

The frequency dependence of the relaxation rate proves the existence of a wide distribution of effective correlation times related to the non-equivalent local barriers [2.67].

After a sudden temperature jump, typically from room temperature to 77 K, a metastability effect is observed in the recovery of the nuclear magnetization [2.68]. This metastability is concentration dependent and may be associated with a slow evolution towards the equilibrium of the dynamical local response function within the different clusters (Panel 2.33). This phenomenon seems to have an equivalent in magnetic spin glasses [2.69].

On cooling below ~ 10 K a progressive changeover in the dependence of the relaxation rate is observed [2.67]. For $T \simeq 4.2$ K and $x = 0.18$ the ω_L-dependence of T_1 displays a behavior indicative of a spectral density for the Na motion of resonant character (Panels 2.33 and 2.4). This spectral density has been tentatively ascribed to Na resonant tunneling between the minima of the local potential, taking into account that in disordered systems one expects a distribution of potential barriers for the interacting units with some of them becoming very large.

The possibility of resonant tunneling of large off-center ions inducing nuclear spin-lattice relaxation had already been suggested in connection with Rb and Cl

---

PANEL 2.32 _____

Quadrupolar Relaxation Study of Disordering Effects and Local Dynamics in KTaO₃:Na

- The ^{23}Na NMR relaxation rate, driven by the time-dependent part of the quadrupolar interaction in KTaO₃:Na shows, for $T \gtrsim 150$ K, a T^2 behavior with no ω_L-dependence (Fig. 2.32a). The extra contribution arising below ~ 150 K, can be explained in terms of a 1D model for the motion of correlated particles in a double-well potential, yielding a spectral density with a low-frequency "central-peak" (Panel 2.2)

$$J(\omega) = 2\tau_{\text{eff}} / \left(1 + \omega^2 \tau_{\text{eff}}^2\right)$$

as depicted in Fig. 2.32b.

- The effective correlation times τ_{eff} for the Na motion are widely distributed as demonstrated by the dependence of the ^{23}Na relaxation rate on the measuring frequency (Fig. 2.32c).

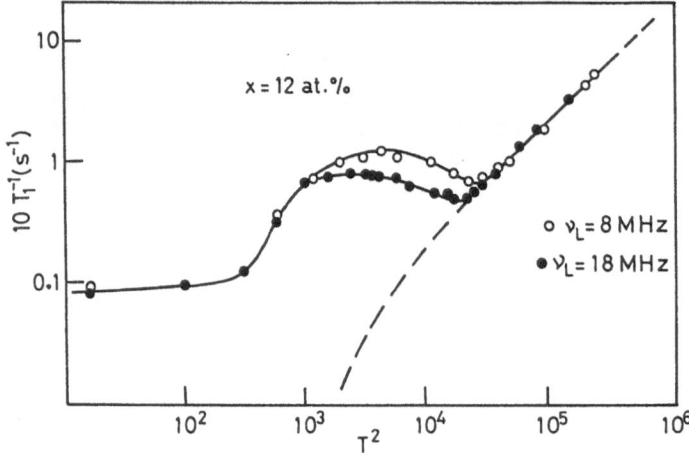

Fig. 2.32a. ^{23}Na relaxation rate in $K_{0.88}Na_{0.12}TaO_3$ as a function of T^2 for two measuring frequencies [2.65]

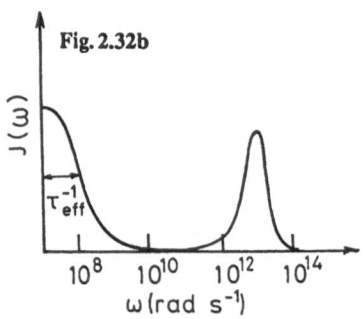

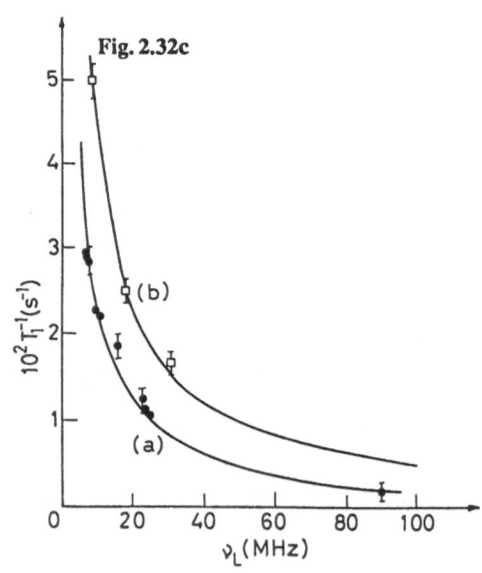

Fig. 2.32b. Spectral density (as in Panel 2.2) used to interpret the extra contribution arising below $T \sim 150$ K in the data of Fig. 2.32a

Fig. 2.32c. Frequency dependence of the ^{23}Na relaxation rate at $T = 77$ K for $x = 0.12$ (•) and $x = 0.05$ (□). The solid lines are the theoretical behavior ($T_1^{-1} \propto \omega_L^{-1}$) expected for a wide distribution of τ_{eff} [2.67]

Quadrupolar Relaxation Study of Metastability and Tunneling Effects in KTaO₃:Na

- Following a temperature jump from RT to below the critical temperature where the local disorder dynamics of the Na atoms sets in, (Panel 2.32) a non-exponential recovery of the nuclear magnetization is observed (Fig. 2.33a). This non-exponential recovery occurs in the intermediate time-interval, when different parts of the crystal have a non-equilibrium local dynamical susceptibility. One has to assume, in this situation, that the spin-spin diffusion is not fast enough to average the different relaxation rates of the different clusters. This is corroborated by the fact that the phenomenon is observed only with low Na concentration, when the different clusters are farther apart.
- The cross over of the Na dynamics from diffusive to resonant below $\sim 10\,\mathrm{K}$ is detected from the frequency dependence of the relaxation rate. While at $T = 77\,\mathrm{K}$, one has the monotonic frequency dependence shown in Fig. 2.32c, at $T \simeq 4.2\,\mathrm{K}$ one observes the ν_L dependence reported in Fig. 2.33b.

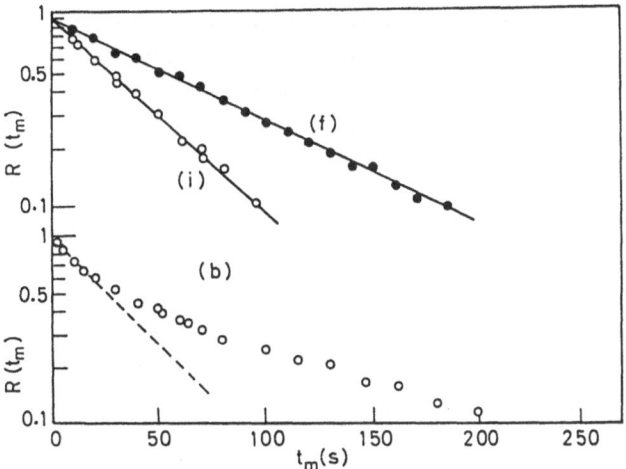

Fig. 2.33a. Effect of metastability observed in the NMR experiment from the evolution, after a sudden temperature jump, of the recovery law of the nuclear magnetization: (*i*) just after the temperature jump; (*f*) final condition, after about 200 h; (*b*) intermediate metastable situation

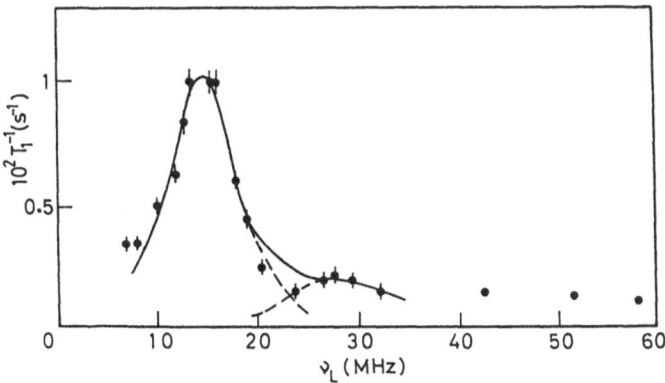

Fig. 2.33b.
Spectralization of the ^{23}Na relaxation rate in $K_{0.82}\,Na_{0.18}TaO_3$ showing the cross over to resonant dynamics. This result yields a spectral density for the Na motion of the form $J(\omega) = \omega_c / [\omega_c^2 + (\omega + \omega_T)^2]$ with a tunneling frequency $\omega_T = 28.5\,\mathrm{MHz}$ and width $\omega_c \simeq \omega_T/4$, the latter being related to a statistical distribution of tunneling rates or to the loss of coherence (*solid line*)

NMR investigation of the dynamic properties of off-center Ag^+ defects in RbCl [2.70].

d) $Rb_{1-x}(ND_4)_xD_2PO_4$. These mixed H-bonded ferroelectric and antiferroelectric crystals have been claimed to represent the closest analog of spin glasses [2.51]. Linewidth and T_1 measurements have evidenced the onset of a wide distribution of correlation times characteristic of the glassy state, also providing information on the distribution of the order parameter [2.52, 71].

The ^{87}Rb and D relaxation rates are driven by the O–D–O intrabond fluctuations through the time-dependent EFGs. The frequency and temperature dependence (Panel 2.34) show a marked departure from the behavior expected for a monodispersive hopping motion (observed for pure RbD_2PO_4). The data are explained on the basis of a correlation function given by a stretched exponential, which should

---

PANEL 2.34

Cluster Dynamics in Disordered Ferroelectric-Antiferroelectric Mixed Crystals (RADP)

- RbH_2PO_4 (RDP) has a continuous ferroelectric (FE) transition near 145 K while $(NH_4)H_2PO_4$ (ADP) becomes antiferroelectric (AFE) below 148 K. The phase diagram (Fig. 2.34a) shows that in RADP mixed crystals with $0.22 \lesssim x \lesssim 0.75$ one has glass-type anomalies: the microscopic details have been studied by H, D, and Rb NMR.

- In marked contrast to the case of $x = 0.78$, the ^{87}Rb and D relaxation rates for $x = 0.35$ and $x = 0.55$ (Fig. 2.34b) can be fitted by an expression of the form

$$T_1^{-1} \propto \int \exp\left[-(t/\tau)^\alpha\right] \exp\left[i\omega_L t \, dt\right] \quad ,$$

with $\alpha = 0.46$ and with an average correlation time, τ, which depends exponentially upon temperature.

- For $x = 0.78$, however, where the ordinary PA→AFE transition occurs, from the ^{87}Rb T_1 one observes the expected slowing-down in the intra-bond motions (Fig. 2.34c).

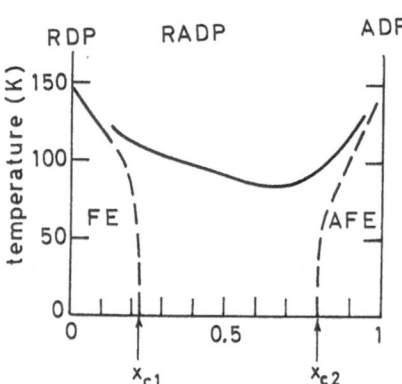

Fig. 2.34a. Phase diagram for the mixed crystals RADP [2.51]

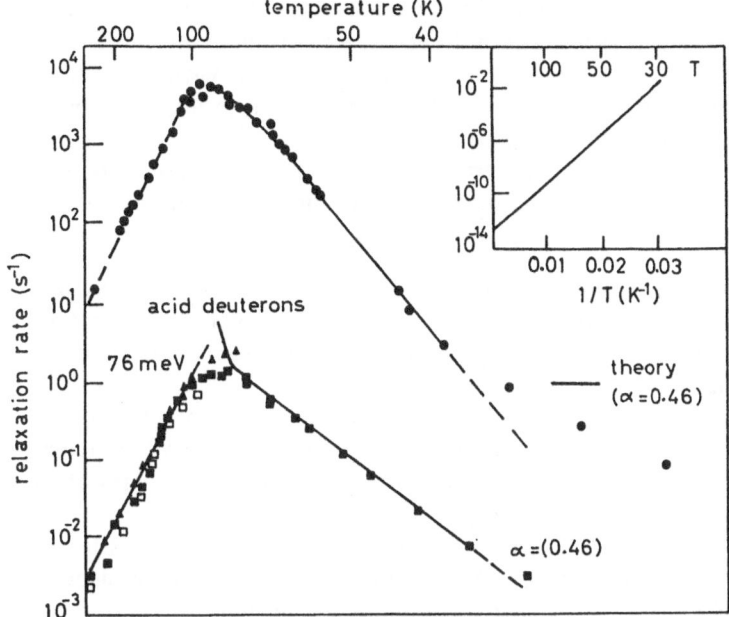

Fig. 2.34b. Temperature behavior of the relaxation rates in RADP and theoretical best-fit lines according to the above equation. From the fits one can derive the average correlation time, which exhibits the dramatic slowing-down shown in the inset

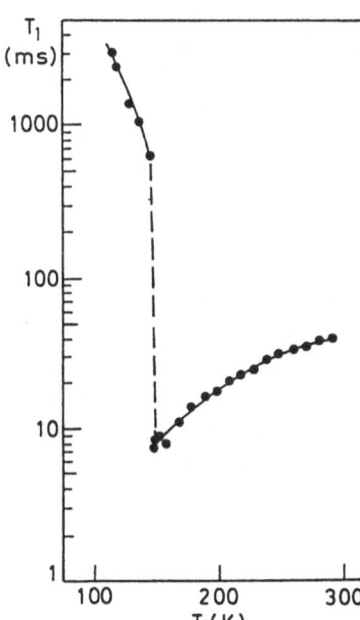

Fig. 2.34c. [87]Rb spin-lattice relaxation times vs T in RADP at $x = 0.78$. A strong discontinuity at T_c is also observed [2.71]

describe heuristically a wide distribution of correlation times. One might observe that a departure from monodispersivity is not, by itself, a clear-cut demonstration of the formation of a glass-type state. However, the $^{87}\text{Rb}(-\frac{1}{2} \leftrightarrow \frac{1}{2})$ NMR line [2.52] shows a line broadening in the temperature region corresponding to the fast-motion condition ($\omega_L \tau \ll 1$), accompanied by a marked asymmetry.

These two features allow one to extract from the data a distribution of static and quasi-static random fields indicative of the freezing-out of the D_2PO_4 groups, which, instead of being sharp as in ordered crystals, is spread over a wide temperature range.

e) **Miscellaneous Systems.** We now mention briefly a few other NMR–NQR and relaxation studies in systems that have some aspects which complete the picture of representative examples in the new challenges arising in disordered crystals compared to the ordered ones.

$(CH_4)_x Kr_{1-x}$ represents a system with orientational disorder in the presence of octupole-octupole interactions. A search for a possible condensation in an octupolar glass phase has been performed through NMR measurements of proton susceptibility, linewidth, T_1, and spin-conversion rate [2.73].

A system similar to $NaCN_{1-x}$–Cl_x, namely $K(CN)_x$–Br_{1-x} has been investigated by ^{13}C and ^{15}N [2.74]. The results cannot be fitted by models in which the molecular reorientations simply slow down with decreasing temperature, thus supporting the hypothesis of a cooperative orientational freezing [2.58, 74].

In isotopically mixed ferroelectric crystals HCl–DCl, ^{35}Cl NQR allowed the measurement of the correlation times, τ_H for the reorientation of the HCl dipoles (where quantum tunneling contributes to the motion) and of τ_D for the DCl dipoles separately [2.75]. The behavior of τ_D as a function of concentration is in qualitative agreement with the theoretical prediction based on the dynamical random mean-field model. In contrast, the values of τ_H exhibit an unexpected persistence in the dilute limit of quantum tunneling (namely, short correlation times) in spite of the coupling with the slow-reorienting DCl dipoles.

Finally it should be mentioned that quadrupole spin-lattice relaxation has been used to gain information about disorder modes in solids in the amorphous state. Here the T_1 of a given nucleus is found to be orders of magnitude shorter than that for the same material in the crystalline state. The results have been accounted for [2.76] in terms of a direct relaxation process (Panel 2.14) involving the quadrupole interaction of the nucleus with the two-level states TLS introduced to explain the anomalous low-temperature thermal properties of glassy materials. It should be noticed that in glasses where the concentration is high these TLSs should interact cooperatively at low temperatures, giving rise to dynamical phenomena similar to the ones described here for systems where the concentration of disordered units is high.

2.5 Incommensurate Phases

NMR–NQR techniques have provided a great amount of information on both static and dynamical properties of incommensurate phases such as the ones generated by atomic displacements from equilibrium positions or by fluctuating dipoles whose statistical average has a spatial variation incommensurate with the underlying lattice. This field has been exhaustively reviewed by *Blinc* [2.33]. Furthermore, incommensurate phases are formed in conducting materials displaying charge density waves (CDW) and in layers of atoms, dipoles or quadrupoles intercalated or adsorbed on surfaces. Recent relevant NMR experiments will be reviewed. The crucial topics for incommensurate phases are the microscopic modulation of the local critical variable and the novel excitations together with their temperature dependence. Basically the first type of information is contained in the resonance spectrum, while the dynamical information is provided by the analysis of the relaxation rates.

2.5.1 Modulation of Local Order Parameter and Phase Transitions

As a consequence of static perturbative effects on NMR levels, or of a distribution of EFGs for NQR, the resonance frequency is related to the local order parameter or "displacement field", establishing a link between the shape of the resonance spectrum $f(\nu)$ and the spatial distribution of the order parameter.

In order to give an illustration we show in Panel 2.35 the linear and quadratic one-dimensional cases. In this illustration we refer to the case where the resonance frequency on a given resonant nucleus depends only on the incommensurate displacement of the resonant nucleus and/or the displacements of adjacent atoms moving in phase with it (so called "local case"; see [2.77]). Experimental examples of this case will be given by referring to $NaNO_2$ and Rb_2ZnCl_4. When the shift of the resonance frequency from the value in the paraelectric phase depends on the displacements of two or more atoms having different phase, one has the "non-local" case. The singularities in the resonance spectra depend here on the phase difference and show a temperature behavior different from the local case. Non-local effects have been observed in the ^{14}N quadrupole-perturbed NMR spectrum of $N(CH_3)_4ZnCl_4$ [2.78].

a) **Dielectric Crystals.** The uniaxial ferroelectric crystal $NaNO_2$ can be represented, approximately, as a one-dimensional chain of permanent NO_2 dipoles, each having two equilibrium positions parallel and antiparallel to the polar axis.

In a narrow temperature interval this crystal displays an incommensurate phase which separates the paraelectric from the ferroelectric phase. The study of this phase has been carried out both by ^{14}N NQR [2.79,80] and by quadrupole perturbed ^{23}Na NMR [2.81]. The ^{14}N NQR resonance frequency can be related to the local order parameter for a NO_2 dipole $\langle s \rangle$, s being a dichotomic order-disorder variable. In the paraelectric phase $\langle s \rangle = 0$ and the temperature dependence of the NQR frequency is related to oscillations of the dipole around the equilibrium position. In the ferroelectric phase ($T \lesssim 164°C$) ν_R is also sensitive to the temperature dependence of the order parameter $\langle s \rangle$. In a narrow temperature interval $164.75 < T < 166°C$ the ^{14}N NQR line indicates a distribution of ν_R, which gives evidence for the distribution of EFG typical of an incommensurate phase. The analysis of the data [2.80] allows one to apply the plane-wave model to the observed incommensurate phase (Panel 2.36).

A system in which the incommensurate phase exists over a large temperature interval and where quadrupole-perturbed NMR spectra have shown the most dramatic

NMR Line-Shape in Incommensurate Phases

• The distribution of the NMR–NQR frequencies is given by

$$f(\nu) \propto \left(\frac{d\nu}{dx}\right)^{-1} = \left(\frac{\partial \nu}{\partial \eta}\frac{\partial \eta}{\partial x}\right)^{-1}$$

where $\eta(x)$ is the local displacement field. In the linear case

$$\eta(x) \propto \cos\left(q_{\text{inc}}x + \delta\right) \quad,$$

while in the quadratic case

$$\eta^2(x) \propto \cos^2\left(q_{\text{inc}}x + \delta\right) \quad.$$

The resulting line shapes are shown in Fig. 2.35.

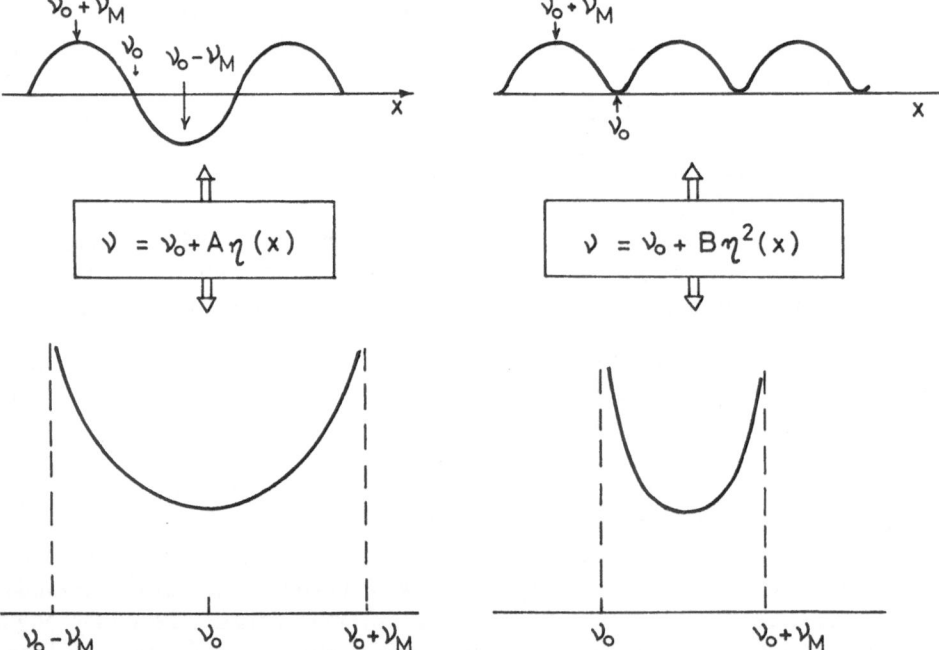

Fig. 2.35. One-dimensional distribution of the local resonance frequency in the linear and the quadratic cases and the corresponding NMR–NQR line shapes

^{14}N NQR in the Incommensurate Phase of Uniaxial Ferroelectric $NaNO_2$

• The relationship between the resonance frequency ν_R and the local order parameter is

$$\nu_R(T) = \nu_0 + B\langle s\rangle^2 + C\langle s\rangle^4 \quad .$$

The line shape exhibits two singularities at $\nu_R' = \nu_0$ and at

$$\nu_R'' = \nu_0 + B\left[s_0(T)\right]^2 \left\{1 + \frac{C}{B}\left[s_0(T)\right]^2\right\} \quad .$$

In agreement with the theory, the temperature dependence of ν_R' is the same as the paraelectric line ν_0, while the singularity at ν_R'' has the temperature dependence of the ferroelectric phase. (Note that the flip frequency of the dipoles is always greater than the resonance frequency.) The resulting NQR line shape, in the various phases, is shown in Fig. 2.36.

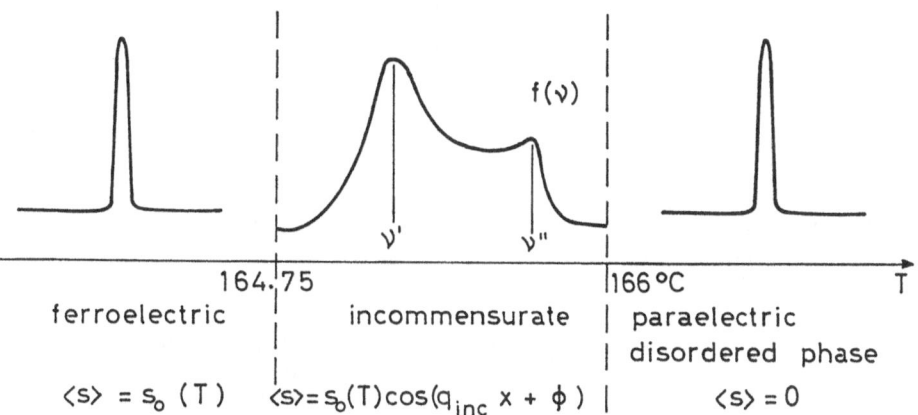

Fig. 2.36. Sketches of the ^{14}N NQR line shapes in the paraelectric, incommensurate and ferroelectric phases of $NaNO_2$. From [2.79, 80]

effects of the onset of the incommensuration, as well as its temperature dependence, is Rb_2ZnCl_4 [2.82, 83]. This crystal undergoes a second-order phase transition from the paraelectric (P) phase to the incommensurate (IC) phase at $T \simeq 30°C$ and a lock-in transition to a ferroelectric phase at $T \simeq -80°C$. The rotation – and distortion – of the $ZnCl_4$ ions as well as the displacements of the Rb nuclei occurring at the onset of the IC phase cause second-order quadrupole effects in the ^{87}Rb NMR spectra.

At the P–IC transition the sharp ^{87}Rb NMR line broadens as illustrated in Panel 2.37. Here the plane wave modulation applies and the second-order quadrupole shift of the central line is a linear function of the nuclear displacements, i.e. of

^{87}Rb NMR of the Incommensurate Phase in Rb_2ZnCl_4

- At the paraelectric-incommensurate phase transition ($T_I \simeq 30°C$) the Rb NMR line becomes inhomogeneously broadened. Since the distance between the singularities in Fig. 2.37a is propotional to the order parameter, one derives $\eta(T) \propto (T_I - T)^{0.36}$.
- At the incommensurate-ferroelectric transition ($T_c \simeq -80°C$) two lines originating from nuclei having a local surrounding typical of the ferroelectric phase appear (Fig. 2.37b). Their integrated intensity $I(T)$ yields a measure of the soliton density $n_s(T)$ (number of solitons per atomic site) which is inversely proportional to the distance between commensurate regions:

$$n_s(T) = 1 - I(T)/I(T < T_c)$$

thus one derives the temperature behavior shown in Fig. 2.37b.

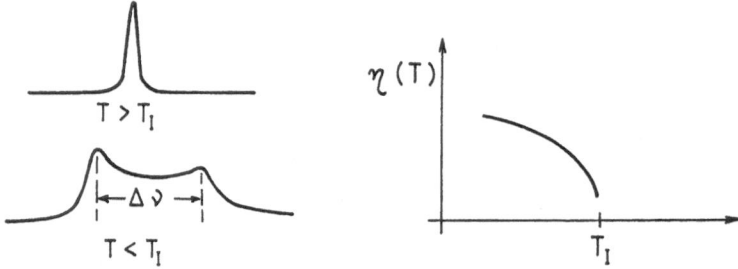

Fig. 2.37a. Sketches of the ^{87}Rb NMR line shapes in Rb_2ZnCl_4 and of the temperature behavior of the order parameter $\eta(T)$ vs temperature

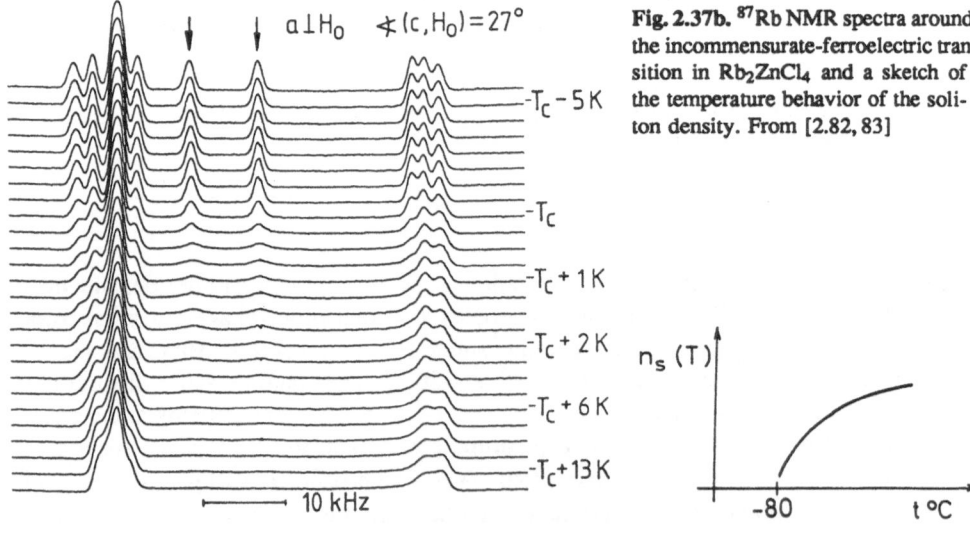

Fig. 2.37b. ^{87}Rb NMR spectra around the incommensurate-ferroelectric transition in Rb_2ZnCl_4 and a sketch of the temperature behavior of the soliton density. From [2.82, 83]

the order parameter. From the frequency separation of the two edge singularities the temperature variation of the amplitude of the incommensurate modulation is derived: the order parameter varies as $(T - T_{\mathrm{I}})^{\beta}$, with critical exponent $\beta \simeq 0.36$, in agreement also with EPR measurements [2.84].

On approaching the lock-in transition to the commensurate ferroelectric phase, additional lines are observed. These lines can be ascribed to commensurate regions resulting from a soliton-like profile of the local order parameter (Panel 2.7). From the integrated intensity of the "commensurate" lines one can derive the temperature dependence of the soliton density $n_s(T)$ (Panel 2.37). The way in which $n_s(T)$ approaches zero for $T \to T_c$ is related to the order of the transition and to the possible existence of a chaotic metastable state with a random pinning of solitons by impurities.

b) Metals. In conducting systems periodic lattice distortions can occur as a consequence of instabilities in the charge density of the conduction electrons (CDWs). The atomic distortion of the lattice and the accompanying charge density wave normally have a q-vector incommensurate with the underlying lattice because it is associated with a divergence of the electronic susceptibility for a particular vector in the reciprocal space, determined by a peculiar topology of the Fermi surface. The normal-to-incommensurate phase transition is usually second order. By lowering the temperature, the elastic energy terms in most cases drive the system into a commensurate phase through a first-order phase transition. Phase transitions of this type are often observed in low-dimensional metals, mainly layered compounds, and one-dimensional, mostly organic, conductors.

From the NMR point of view, the CDW incommensurate phases may be studied by two probes: the quadrupole perturbation of the spectrum and/or the shift in the resonance frequency due to the local magnetic field of the conduction electrons (Knight shift).

In the transition metal dicalcogenide layered compound $2H$–$NbSe_2$ extensive NMR studies have been carried out both on the quadrupolar nucleus ^{93}Nb [2.85] and on the $I = \frac{1}{2}$ Se nucleus [2.86]. The CDW spatially modulates the electron density and thus the EFGs and the Knight shift at the ^{93}Nb site. It is reasonable to assume that both the EFG variation $\Delta q(R)$ and the Knight shift distribution $\Delta K(R)$ are proportional to the CDW $\Delta \varrho(R)$. The atomic displacement and the CDW amplitude $\Delta \varrho_0$ are proportional to the order parameter. Therefore the resonance frequency is linearly related to the order parameter and for the 1D plane wave CDW the expected line shape is the one depicted in Panel 2.35.

For a CDW described by a superposition of three independent plane waves, with different q-values, the line shape consists of two step singularities and an infinite singularity (Panel 2.38).

The NMR investigation of $2H$–$NbSe_2$ gives a line shape in substantial agreement with the 3-plane-wave model. No evidence was found for additional lines as one would expect were the incommensurate state to evolve on approaching the lock-in transition in terms of McMillan's discommensurations separated by commensurate regions (similar to the case illustrated in Panel 2.37).

153

Incommensurate Phase in Metallic NbSe₂ by ⁹³Nb NMR

- The onset of an incommensurate CDW phase is accompanied by drastic changes in the ^{93}Nb NMR line shape. The temperature evolution of the $(-\frac{1}{2}, -\frac{3}{2})$ satellite line is consistent with the onset of a triple incommensurate CDW (Fig. 2.38a).
- The temperature dependence of the distance between the two singularities yields directly the relative order parameter (Fig. 2.38b).

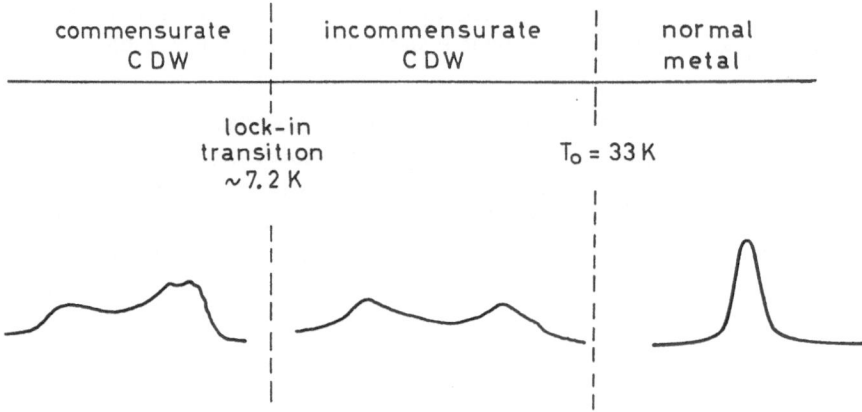

commensurate CDW | incommensurate CDW | normal metal

lock-in transition ~7.2 K

$T_0 = 33$ K

Fig. 2.38a. Evolution of the ^{93}Nb NMR line shape in NbSe₂

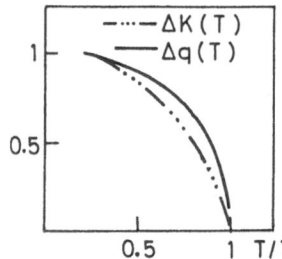

$- \cdots - \Delta K(T)$
$--- \Delta q(T)$

1

0.5

$0.5 \quad 1 \quad T/T_0$

Fig. 2.38b. Temperature behavior of the order parameter for NbSe₂; note that $\Delta\nu(T)/\Delta\nu(4.2) \sim \Delta K(T)/\Delta K(4.2) \sim \Delta q(T)/\Delta q(4.2) \sim \Delta\varrho(T)/\Delta\varrho(4.2)$. From [2.85]

In some instances, incommensurate phases are very complex. For example in NbSe₃, which consists of three chains or columns, each of which is represented twice in the unit cell, incommensurate CDWs appear with different characteristics in the three conducting chains. Here, the ^{93}Nb NQR spectra have been studied and even though their analysis is very complex, they nonetheless provide valuable microscopic information [2.87].

A very interesting effect expected in CDW systems is the possibility of charge transport associated with a collective motion of the electrons, which can be viewed as a rigid sliding motion of CDWs (Frölich mode). Recently, it has been possible to observe this phenomenon directly from the narrowing of the inhomogeneously broadened NMR line in NbSe$_3$ [2.88] and in Rb$_{0.3}$MoO$_3$ [2.89] when an applied electric field exceeds a threshold value.

2.5.2 Excitations in the Incommensurate Phase and Relaxation Rates

The excitations in incommensurate phases produce striking effects on the nuclear relaxation: an anomalously short relaxation time, a dependence on the spectral range of the line irradiated, and a peculiar temperature and frequency dependence. These effects, over a wide temperature range, are essentially due to the persistence in the incommensurate phase of large, low-frequency fluctuations, of the type also present in ordered systems close to T_c (Panel 2.39). In fact, on approaching T_{IC} in the normal phase one usually observes a divergent behavior of T_1^{-1} of the same type found in "ordered" crystals and related to the slowing-down of the critical fluctuations. Only well below the locking temperature T_c in the commensurate phase, do the nuclear relaxation rates resume their normal values (Panel 2.40).

PANEL 2.39 _____

Nuclear Spin-Lattice Relaxation Due to Phasons and Amplitudons

- The EFGs are expanded in a series of atomic displacements

$$V_{JK}(t) = V_{JK}(0) + \sum_i \left(\frac{\partial V_{JK}}{\partial u_i} \right) u_i + \frac{1}{2} \sum_{i,J} \left(\frac{\partial^2 V_{JK}}{\partial u_i \partial u_s} \right) u_i u_s \quad .$$

Direct process   Raman process

- The relaxation rate is related to the spectral densities $J(\omega_L)$ of $\langle u_i(0)u_i(t)\rangle_0$; by Fourier transformation one has:

$$T_1^{-1} \propto \sum_q S(q, \omega_L) \equiv \frac{T}{\omega_L} \sum_q \chi_q''(\omega_L) \quad .$$

- The dynamic susceptibility is approximated by a damped-harmonic oscillator type

$$\chi_q''(\omega) = \Gamma\omega / \left[\omega^2 - \omega^2(q) \right]^2 + \Gamma^2\omega^2 \quad .$$

- The q-summation is performed by using the appropriate dispersion relations [2.90, 91]:

$$
\begin{aligned}
\text{phasons} \qquad & \omega_{\mathrm{ph}}^2 = a\left(q - q_{\mathrm{inc}}\right)^2 \\
\text{amplitudons} \qquad & \omega_{\mathrm{amp}}^2 = b\left(T_I - T\right) + \omega_{\mathrm{ph}}^2 ,
\end{aligned}
$$

or by using other expressions for the multisoliton lattice and for the case when a phase pinning perturbation induces a gap in the phason spectrum; see also Panels 2.14 and 2.19.

Experimental Results on Nuclear Relaxation Driven by Phasons and Amplitudons

- The theoretical expectations for a nuclear spin-lattice relaxation process driven by "solitons" are found to agree with the experimental results of ^{14}N NMR T_1 in $[N(CH_3)_4]_2ZnCl_4$ (Fig. 2.40a).
- The ^{87}Rb relaxation in Rb_2ZnCl_4 is driven by phasons. On approaching the lock-in transition one has the cross-over from phason excitations in the plane-wave limit to those of a multisoliton lattice. The relaxation rate traces the density of phase solitons (Fig. 2.40b).

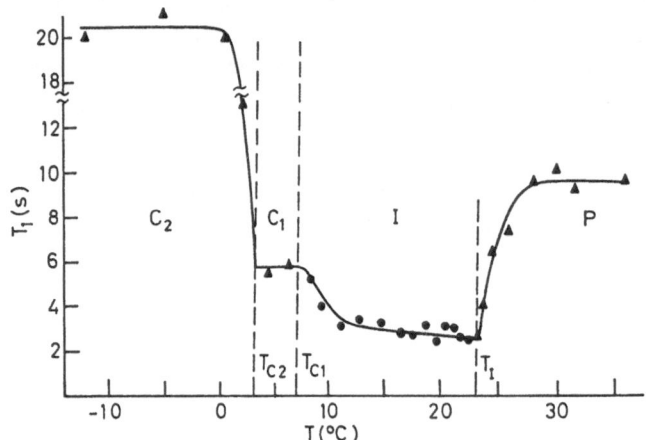

Fig. 2.40a. Temperature behavior of ^{14}N NMR T_1 [2.92]

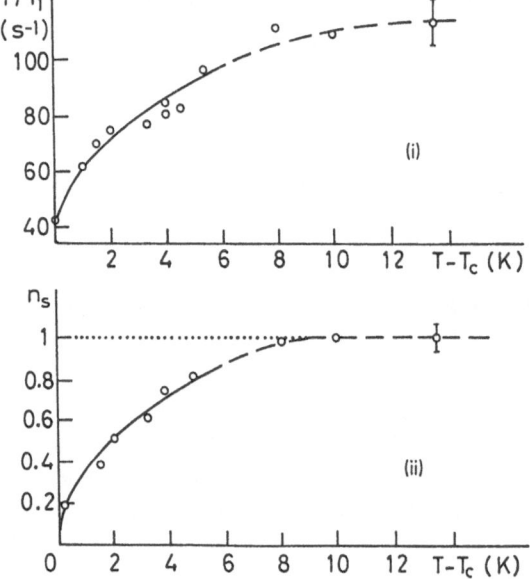

Fig. 2.40b. Temperature behavior of the ^{87}Rb relaxation rates for $T \rightarrow T_c^+$ (lock-in temperature) (*i*) and corresponding density n_s of phase solitons (*ii*) [2.92, 93]

For $T_c < T < T_{IC}$ the relaxation is driven by phason and amplitudon excitations and the theory for the spin-lattice relaxation can be carried out along the same lines as sketched in Sect. 2.3 for the phase transitions in ordered crystals (Panel 2.39).

If, on approaching the commensurate phase at T_c, a cross-over from the plane wave to the multisoliton lattice occurs, one observes a drop in the relaxation rate. This can be understood qualitatively by considering the decrease of the regions of the crystal where the phase modulation associated to the phase solitons induces fluctuations in the EFGs. A crude proportionality of the relaxation rate to the number of solitons is therefore expected (Panel 2.40).

If the bandwidth Δ_{BW} is large enough to irradiate the whole inhomogeneously broadened spectrum, then T_1 is almost T-independent, being dominated by phasons. If their dispersion is gapless, so that a sizeable number of excitations is in the radio-frequency range, T_1 should depend on the linewidth. If one reduces the bandwidth of the irradiation to less than the linewidth then it is possible, in principle, to separate the contribution of amplitudons from that of the phasons by irradiating different sections of the resonance spectrum. In fact, if we refer for simplicity to the linear plane-wave case of Panel 2.35, one can see that the regions of the spectrum around singularities come from nuclei which sit in the maxima (minima) of the plane-wave modulation and are thus sensitive mostly to amplitudons.

Experimental evidence of a T_1 variation across the resonance line has been obtained in the incommensurate phase of $[N(CH_3)_4]_2ZnCl_4$ [2.92] (Fig. 2.41a).

The relaxation rate is a very sensitive tool to detect the possible presence of gapless phasons. In fact the presence of a gap in the phason dispersion spectrum causes a drastic decrease in the density of excitations in the rf range. Thus the relaxation rate is smaller, frequency independent, and inversely proportional to the phason gap. An ω_L-dependence of T_1 in the incommensurate phase has been detected in biphenyl [2.94]. In this molecular crystal, $C_{12}H_{10}$, an order-disorder type phase transition occurs. Above $T_{IC} \sim 40\,K$, the two phenyl groups are in fact coplanar because of fast mutual fluctuations of the two rings in opposite senses. Below T_{IC} the time averaged positions of the two rings are no longer coplanar and the distortion (twist angle) of the unit cells oscillates in space with an incommensurate wave-vector. The proton T_1 can be related to the fluctuations of the dipolar field due to the order-disorder twisting in the normal phase ($T > T_{IC}$), and the phason excitation for $T < T_{IC}$. On approaching T_{IC} from above, a marked divergence of the relaxation rate signals the critical slowing-down. In the incommensurate phase below T_{IC}, T_1^{-1} remains large and frequency dependent, suggesting a kind of persistence of the low frequency fluctuations characteristic of the approach to a structural phase transition. The experimentally observed dependence goes as $T_1^{-1} \propto \omega_L^{-3/2}$, which does not entirely agree with a theoretical prediction for gapless overdamped phasons, yielding $T_1^{-1} \propto \omega_L^{-1/2}$ (Fig. 2.41b).

2.5.3 Incommensurate Phases in Layers: Cs Ions Intercalated in Graphite

A category of incommensurate phases that show a wealth of interesting effects are those originating in the physisorption or intercalation of ions, dipoles or quadrupoles on a substrate. The competition between the interaction energy among the particles

Relaxation Rates in Incommensurate Systems

- In $[N(CH_3)_4]_2ZnCl_4$ the ^{14}N NMR spectrum displays the two singularities of the inhomogeneously broadened line typical of the incommensurate plane-wave situation. The ^{14}N effective spin-lattice relaxation rate varies across the resonance line (Fig. 2.41a).
- In biphenyl $(C_{12}H_{10})$ the proton relaxation rate gives a clear indication of a frequency dependence of T_1 in the incommensurate phase, as expected for a gapless spectrum of low frequency excitations (Fig. 2.41b).

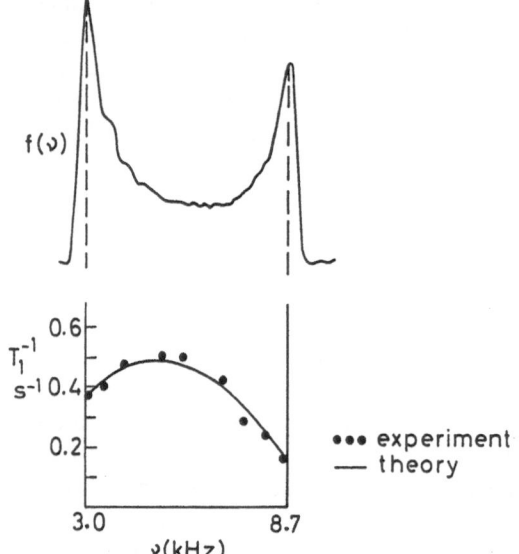

Fig. 2.41a. Experimental evidence of a relaxation rate varying across the NMR line [2.92]

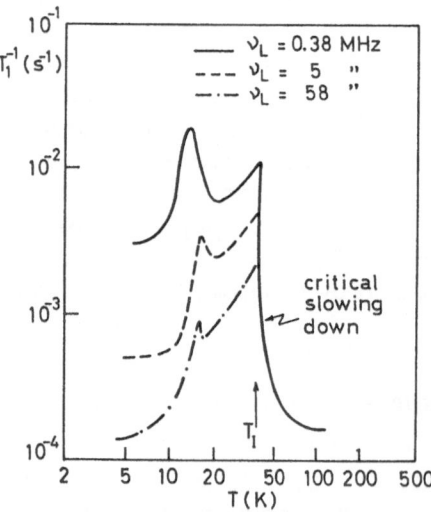

Fig. 2.41b. Proton spin-lattice relaxation rate vs T in $C_{12}H_{10}$, showing the dependence on the measuring frequency. (Sketch from data in [2.94])

intercalated or physisorbed and the periodicity of the potential due to the underlying lattice is the driving mechanism which determines whether the layer is commensurate (registered) or incommensurate (floating). The transition between the two phases can be driven by temperature and/or the concentration of the adsorbed (intercalating) species.

Among the relatively few NMR–NQR studies of incommensurate phases in adsorbates or intercalates (partly due to the poor signal-to-noise ratio) we will briefly describe the system of Cs ions intercalated between the carbon layers of pyrolytic graphite. A great advantage of intercalates is the possibility of controlling the concentration of the intercalant so as to obtain either the commensurate or the incommensurate system (stages) while still preserving the monolayer character [2.95]. For CsC_8, for example, the Cs ions occupy the positions corresponding to the center of the carbon hexagons and therefore the monolayer phase is commensurate. For CsC_{24}, on the other hand, the areal density is reduced to $\frac{2}{3}$ allowing for a triangular lattice with an average Cs–Cs distance which is no longer commensurate with the underlying carbon lattice. In this situation one has the interesting circumstance of an incommensurate phase which has a wealth of unusual excitations over a very wide temperature range.

^{133}Cs quadrupole-perturbed NMR spectra and quadrupolar spin-lattice relaxation rate have yielded information on the short-range order of the in-plane structure and on the low-frequency excitations [2.96]. Over the whole temperature range, from liquid helium temperature up to the deintercalation temperature of $\sim 640\,\mathrm{K}$, a fully resolved first-order quadrupole split NMR spectrum is observed when the external magnetic field is in the plane of the layer. The anomalous angular dependence of the spectrum and the marked decrease in intensity with increasing temperature support the picture of incommensurate regions at the boundaries of short-range ordered commensurate islands [2.97].

The measured ^{133}Cs relaxation rate turns out to be four orders of magnitude larger than that induced by ordinary phonons in Cs halides. Since the Korringa mechanism (relaxation due to conduction electrons) and an in-plane liquid-like diffusion mechanism can be ruled out on the grounds of order of magnitude and temperature and frequency dependence, it has been possible to ascribe the very effective relaxation mechanism to the presence on phonon-like low-frequency modes. These modes correspond to the sliding optical modes observed in Raman scattering and are determined by the in-plane shear vibration of Cs clusters against the two carbon bounding layers. Since the dispersion is small, the normalized distribution function describing the density of phonon states is broadened by the inverse of the lifetime of the clusters of correlated Cs, the lifetime being limited by the diffusion process. The effective frequency for these modes can be derived from T_1 data and the departure of T_1^{-1} from the T^2 dependence is used to infer a softening of this frequency on approaching the deintercalation temperature (Panel 2.42).

^{133}Cs *NMR in Intercalated Graphite*

- The NMR quadrupole-perturbed spectrum of Cs intercalated in graphite (CsC_{24}), with the first-order splitting of the satellite lines ($I = \frac{7}{2}$), indicates an EFG tensor of axial symmetry present, on the average, at the Cs site. The presence of discommensurations (apparent in the projection of the Cs positions on the basal plane containing the carbon hexagons) causes a local distribution of the directions of V_{zz} (with respect to the normal to the basal plane) and a departure from axial symmetry ($\eta \neq 0$). This distribution produces the disappearance of the satellite lines when H_0 goes out of the basal plane (Fig. 2.42a).

(i)

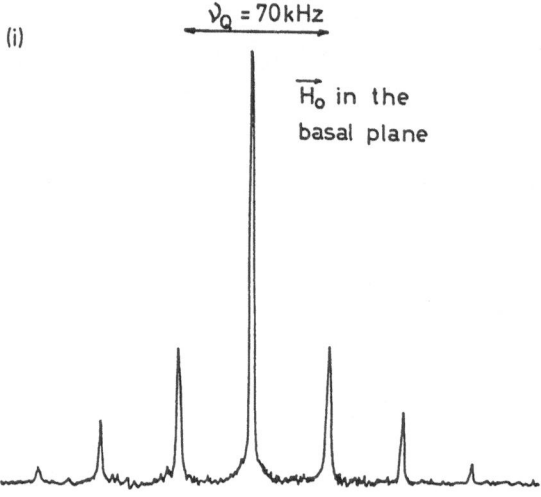

(ii)

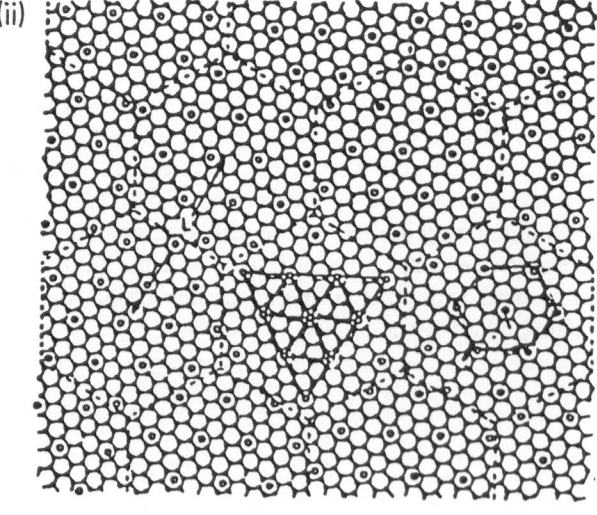

Fig. 2.42a. ^{133}Cs NMR quadrupole-perturbed spectrum in CsC_{24} (*i*). The discommensurations causing the disappearance of the satellite lines are shown in (*ii*)

- The ^{133}Cs relaxation rates (Fig. 2.42b) can be compared with those expected for the cases:

  i) ordinary lattice phonons (Raman process) as, for instance, in CsCl; then

  $$T_1^{-1} \sim 10^{-8}\, T^2\, \mathrm{s}^{-1} \quad ;\text{(Panel 2.14)}$$

  ii) low frequency phonon-like sliding modes (interrupted by diffusion); then

  $$T_1^{-1} = BT^2/\omega_{\mathrm{E}}^4\, \Gamma \sim 10^{-4}\, T^2\, \mathrm{s}^{-1}$$

  where the frequency of the sliding mode is $\omega_{\mathrm{E}} = 10\,\mathrm{cm}^{-1}$ and the width due to the overdamping associated with the diffusion is

  $\Gamma \sim 3 \times 10^{11}\,\mathrm{rad\,s}^{-1}$.

  The experimental findings strongly support case (ii).

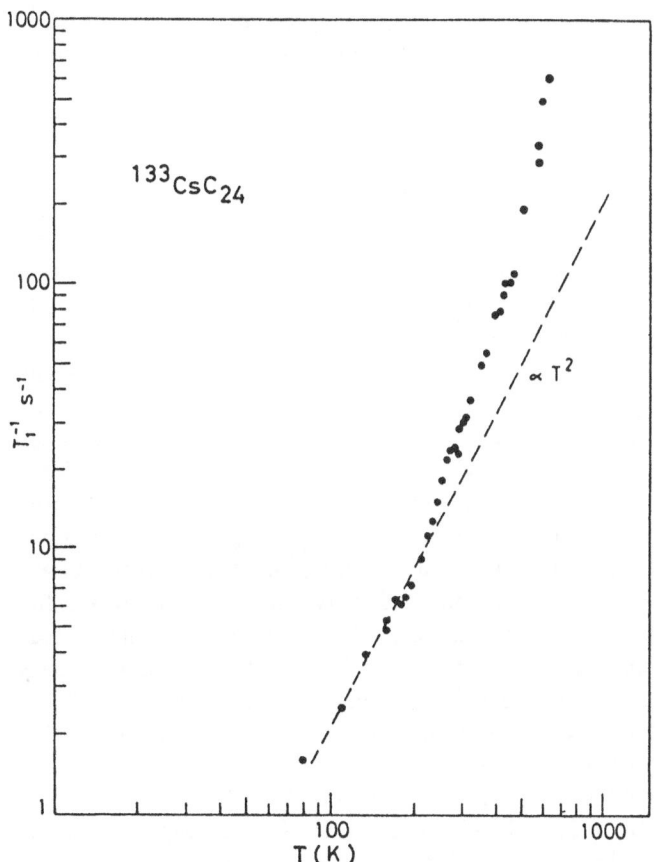

Fig. 2.42b. ^{133}Cs NMR relaxation rates in CsC$_{24}$ as a function of temperature [2.96]

2.6 Non-linear Phenomena, Central Peak and Pretransitional Clusters

In this section we discuss the NMR–NQR experiments related to the occurrence of strong non-linearity. Small non-linearities causing renormalization of dynamical quantities, damping, broadening of spectral densities, etc., are not included here (see Sect. 2.2). In the context of SPTs, the non-linearity can appear in the neighborhood of T_c, where large fluctuations cause the breakdown of MFA and the insurgence of a "disorder" in the dynamics. These dynamics do not necessarily comply with the symmetry imposed by the condition of small displacements for displacive transitions but they appear in higher order approximations for order-disorder transitions.

NMR and NQR are, in a way, the natural experimental tools to detect non-linearities since these show up in central peaks, and also to take a "snap shot" of pretransitional clusters anticipating the low-temperature phase. Furthermore, tunneling and domain wall and/or soliton motions have dramatic effects on relaxation. In crystals with randomly distributed impurities, non-linear dynamics can be generated locally around the impurities themselves, giving rise to phenomena, such as central peaks and local freezing, which are extrinsic non-linear effects and which lead to similar manifestations in NMR–NQR experiments.

2.6.1 Central Peak

As discussed in Sect. 2.1.4, the CP is a narrow component, centered at zero frequency, which is superimposed on, and rises above the mean-field dynamical structure factor described by lattice soft-modes in displacive PTs and by soft pseudo-spin or diffusional modes in order-disorder PTs.

The various nuclear relaxation times (T_1, $T_{1\varrho}$, T_{1D}, T_2) probe the spectral density of the motions typically in the frequency range 0–100 MHz. An "a-priori" problem in the correct interpretation of the data is the need to discriminate between low-frequency contributions due to CP mechanisms and those related to ordinary mean-field slowing-down. The criteria to ensure that one is detecting a CP could be: (i) a full spectralization, from the frequency dependence (ω_0, ω_1, ω_L) of the relaxation rate, indicating the presence of a narrow component *in addition* to the possible overdamped soft mode or order-disorder diffusive mode; (ii) when the full spectralization is not possible, one has to observe a relaxation rate much larger than that expected for normal phonon relaxation and indicating a characteristic frequency Γ_c orders of magnitude smaller than the soft frequency ω_s; (iii) for $T \to T_c^+$ the enhancement of the relaxation rate due to slowing-down of Γ_c must be characterized by critical exponents different from the mean-field values.

Among the very few examples of T_1 spectralization of CP we mention the proton and deuteron relaxation results in KDP [2.98]. The proton T_1 is strongly magnetic-field dependent for $T \to T_c^+$ (Panel 2.43), indicating the presence of a CP whose width is of the order of MHz. As a matter of fact, the width of the CP becomes narrower than the Larmor frequency thus leading to a decrease of T_1 for $T \to T_c^+$. The deuteron relaxation rate (in deuterated KDP) does not show any field dependence

Relaxation Effects from Central Peak in Order-Disorder Crystals

- Since the local spectral density at the measuring frequency ν_L is $J(\omega_L) = \sum_{\vec{q}} S(\vec{q}, \omega_L)$, in principle, by varying ω_L one can spectralize the CP (Fig. 2.43a).
- In KH_2PO_4 the temperature dependence of the proton relaxation time indicates a CP of width $\Gamma_c \sim 10\,MHz$ at $T \simeq 150\,K$. In contrast, the deuteron relaxation time reflects the slowing-down of the soft mode frequency ω_s (Fig. 2.43b).
- By varying the magnetic field, the dependence of the proton T_1 in KH_2PO_4 (Fig. 2.43c (i)), indicates a narrow CP, while the independence (Fig. 2.43c (ii)) of the deuteron T_1 suggests that in this frequency range ($\omega_L = \gamma H$) the spectral density is flat in $K_2D_2PO_4$ (Fig. 2.43a).

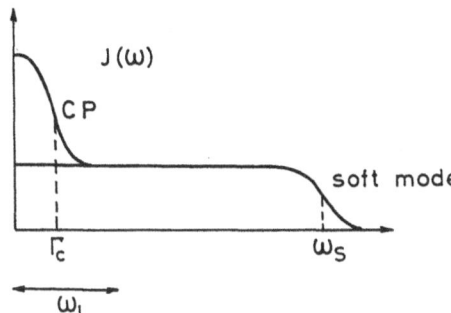

Fig. 2.43a. Sketch of a local spectral density with a central peak and a soft mode of relaxational character

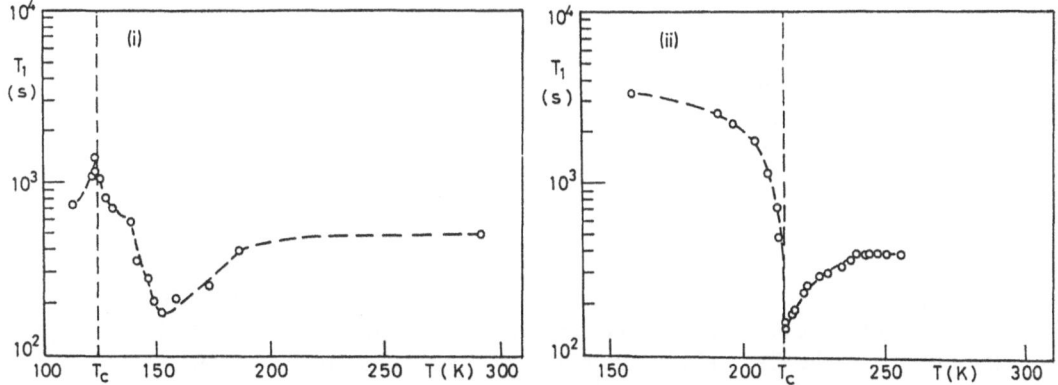

Fig. 2.43b. Proton relaxation time in KH_2PO_4 vs T, displaying a minimum when $\Gamma_c \sim \omega_L$ (10.75 MHz) (*i*); the deuteron T_1 (*ii*) decreases critically for $T \rightarrow T_c^+$ reflecting the soft-mode behavior [2.98] (Panel 2.19)

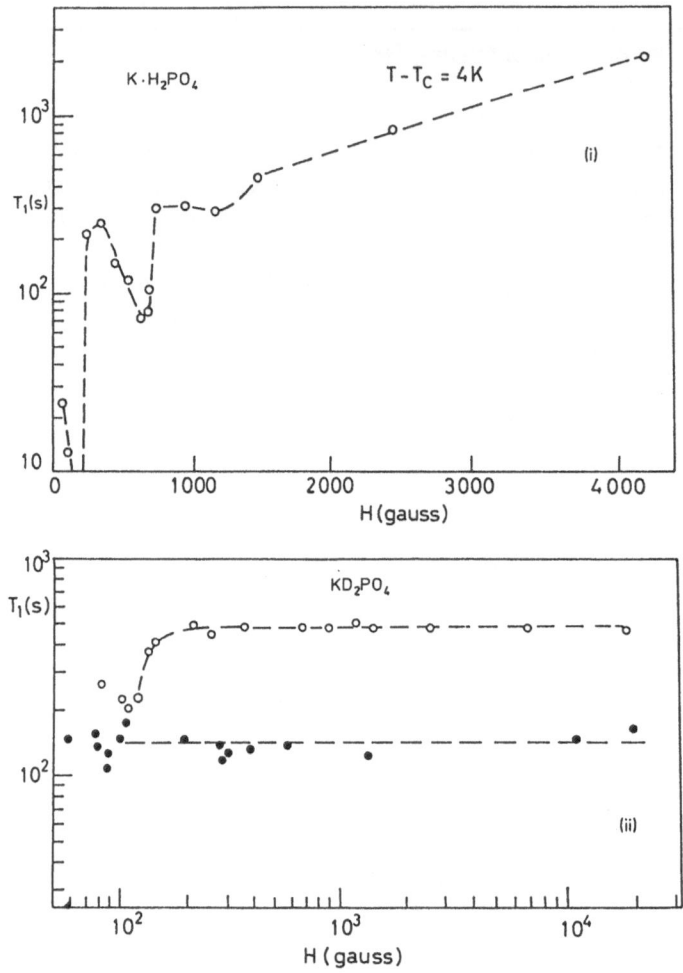

Fig. 2.43c. Proton (*i*) and deuteron (*ii*) T_1 in KH_2PO_4 as a function of the magnetic field [2.98]

near T_c and displays a continuous increase for $T \rightarrow T_c^+$. This result is consistent with an almost flat spectral density in the presence of the usual mean field slowing-down, indicating that the CP is either two orders of magnitude narrower than in KH_2PO_4 or that it is entirely absent (or very broad).

An example of a CP detection without a complete spectralization is offered by the T_1 and $T_{1\varrho}$ study of ^{23}Na in $NaNbO_3$ around the displacive antiferrodistortive transition at $T \simeq 641°C$ [2.99]. The relaxation rate is practically constant for $T \gtrsim T_c + 70°C$, at variance to the T^2 temperature dependence expected for a Raman indirect process due to ordinary lattice phonons (Sect. 2.2). For $T \rightarrow T_c^+$ one observes a divergent behavior of T_1; for reduced temperature $\varepsilon \simeq 5 \times 10^{-3}$ a departure from the divergent behavior is observed as expected for $\Gamma_c \sim \omega_L$ (a condition supported by $T_{1\varrho}$ measurements also) (Panel 2.44). Furthermore, the indirect estimate of Γ_c

Central Peak in Displacive Crystals: NaNbO₃

- The ^{23}Na relaxation rate T_1^{-1} shows a departure from the theoretical behavior $T_1^{-1} \propto \sum_{\bar{q}} \langle \varphi_{\bar{q}}^2 \rangle / \Gamma_q$ (curve *a* in Fig. 2.44) that would be expected if the frequency

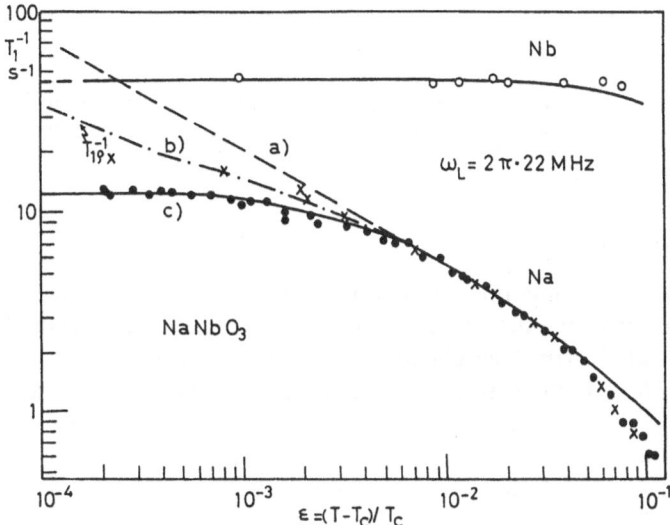

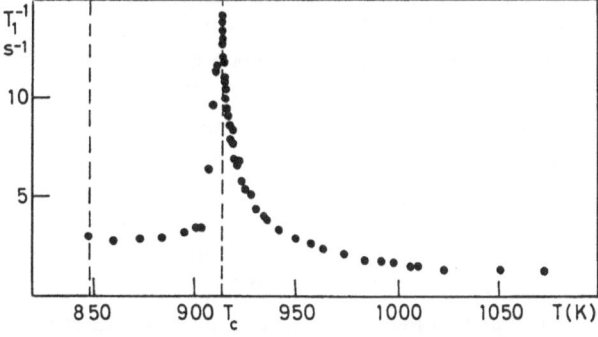

Fig. 2.44. ^{23}Na relaxation rate in NaNbO₃ vs temperature on approaching the cubic-tetragonal PT: (•)T_1^{-1}; (x)$T_{1\varrho}^{-1}$. The divergent behavior of the relaxation rate vs reduced temperature allows one to estimate indirectly the width of the CP

width Γ_c of the CP were to stay always greater than ω_L.

Since the rotating frame relaxation rate

$$T_{1\varrho}^{-1} \simeq \tfrac{7}{10} T_1^{-1} + A J(2\omega_1)$$

also probes the spectral density at $\omega_1 \sim 2\pi \times 20\,\text{kHz}$, a divergent behavior is observed even for $\varepsilon \lesssim 5 \times 10^{-3}$ where $\omega_1 \lesssim \Gamma_c \lesssim \omega_L$ (curve *b*).

Curve *c* is the theoretical behavior expected for the relaxation rate T_1^{-1} when one allows for the width of the CP to become of the order of ω_L. From the fit of the T_1^{-1} data one has:

$$\Gamma_c(T) \sim 2 \times 10^{11} \left[(T - T_c)/T_c \right]^{1.2} \quad .$$

(with $\Gamma_c \sim 2\omega_L \sim 280\,\text{MHz}$ at $T \simeq T_c + 4\,\text{K}$) and

$$\langle \varphi^2 \rangle = \frac{1}{N} \sum_q \langle \varphi_q^2 \rangle \simeq (1.7°)^2 \ .$$

---

around T_c yields a value ($\sim 200\,\text{MHz}$) several orders of magnitude smaller than the frequencies of overdamped soft modes, with a divergence of the form $\Gamma_c \propto (T - T_c)^\gamma$, with critical exponent $\gamma \simeq 1.2$.

2.6.2 Pretransitional Phenomena

Phenomena associated with strong non-linearity that have been evidenced in NMR–NQR studies in ways other than the detection of a CP in the dynamical structural factor, are the pretransitional clusters. The phenomenon mainly consists of the formation of clusters of one phase in the matrix of the other, e.g., ferroelectric clusters in a paraelectric matrix or simultaneous presence of tetragonal and cubic phases. NMR and NQR experiments detect these clusters by extra lines in the spectra, by broadening, or by the pretransitional progressive disappearance of signals.

a) RbCaF$_3$ – a 2D FT NMR Study. A direct observation of the coexistence of cubic and tetragonal phases (both above and below T_c) has been reported in RbCaF$_3$ by means of a 2D Fourier transform NMR method [2.100]. The Rb spectrum $g(f_2, t_1)$ is recorded for an excitation rf pulse of different lengths t_1. Then a FT over t_1 leads to a 2D representation in terms of the two frequencies f_2 and f_1. This allows one to separate the signal coming from regions in the sample where tetragonal distortions smear out the satellite transitions thus reducing the pulse length necessary to achieve a $\pi/2$ pulse for the fictitious spin corresponding to the central line $(\pm\frac{1}{2})$ transition (Panel 2.45). It remains to be decided whether the pretransitional clusters are of intrinsic or extrinsic origin. It should be noted that measurements performed on a different sample have shown an enhancement in T_1^{-1} for $T \to T_c^+$, which can be associated with the presence of a CP, but no change in signal intensity and/or shifts

---

PANEL 2.45 _____

NMR Evidence for Pretransitional Clusters

- The ^{87}Rb 2D NMR spectrum in RbCaF$_3$ (more properly 2D solid state nutation NMR) at $T \simeq T_c = 195.5\,\text{K}$ (Fig. 2.45a) shows the appearance of a second maximum in the signal for pulse lengths of about half of those characteristic for the cubic structure. This indicates the simultaneous presence of cubic clusters and of tetragonal clusters for which only the central line $(\pm\frac{1}{2})$ is irradiated (data from [2.100]).

- In squaric acid the ^{13}C high-resolution spectrum shows, for $T \simeq T_c = 370\,\text{K}$, the coexistence of the low temperature phase (four lines) and of the high temperature phase (two lines). The relative number of nuclei present in the two phases is measured by the integrated intensity, I_L and I_H, of the lines corresponding to the low temperature and high temperature phases respectively (Fig. 2.45b).

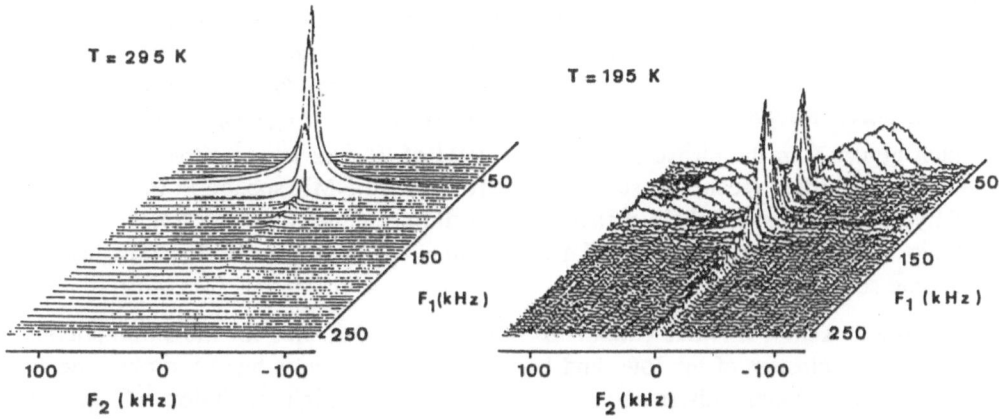

Fig. 2.45a. The 2D ^{87}Rb NMR spectrum in RbCaF$_3$ close to T_c compared with the high temperature spectrum [2.100]

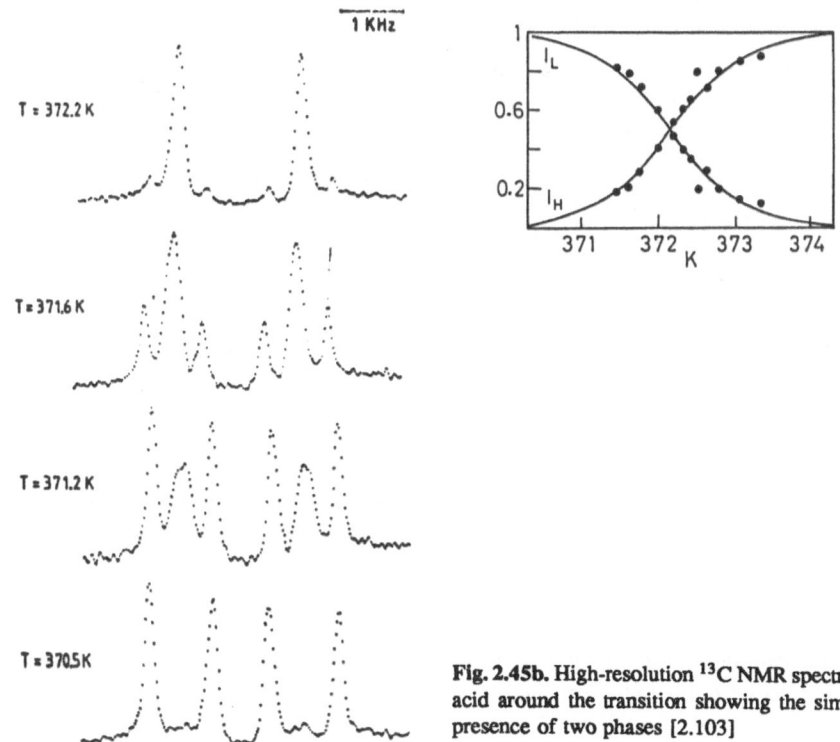

Fig. 2.45b. High-resolution ^{13}C NMR spectra in squaric acid around the transition showing the simultaneous presence of two phases [2.103]

[2.41, 101]. Thus it would appear that the observed pretransitional clusters are sample dependent and must therefore be ascribed to strains and/or impurities. However, in order to establish this point one should perform a detailed study of both spin-lattice relaxation and NMR on the same sample and obtain the characteristic frequency of the CP to be compared with the life-time of the tetragonal clusters. In fact, the smearing of the satellites observed in $RbCaF_3$ implies a lifetime of the pretransitional clusters of the order of 10^{-6} s, as derived from the quadrupole coupling constant. It should be pointed out that if the clusters were to be considered of intrinsic nature for a second-order transition, one would expect a spin-lattice relaxation rate for $T > T_c$ which goes through a maximum of the order of $T_1^{-1} \sim \langle \omega_Q^2 \rangle / \omega_L \simeq 10^3\text{-}10^4\,s^{-1}$, much higher than the measured one [2.41, 101] of $T_1^{-1} \simeq 30\,s^{-1}$. One should also mention that more recent measurements [2.102] of Rb NMR spectra in the presence of hydrostatic pressure show that the pretransitional clusters are no longer present above T_c for a pressure around 5.4 kbar, which, in turn, causes the transition to become second order.

b) Squaric Acid – a NMR Chemical Shift Study. Evidence for the simultaneous existence of clusters of the low- and high-temperature phases has been obtained in the two-dimensional antiferroelectric squaric acid, from high resolution ^{13}C spectra [2.103] (Panel 2.18). In a temperature range of about 2°C around the transition, both the four lines characteristic of the chemical shift tensor at low temperature and the two lines for the high temperature phase are observed, with relative intensities varying continuously (Panel 2.45). The frequency shift $\Delta \nu$ being of the order of 0.5 KHz, the lifetime of the clusters should be $\tau > \Delta \nu^{-1} \simeq 2 \times 10^{-3}$ s. These quasi-static clusters are found to depend on the impurity content and thus are believed to be of extrinsic character.

c) K_2OsCl_6 – a NQR Line-Shape Study. The ^{35}Cl NQR line in the antifluorite crystal K_2OsCl_6 (Panel 2.17) shows the appearance of a broad component shifted with respect to the NQR line of the cubic phase as $T \rightarrow T_c^+$ (Panel 2.46).

Correspondingly, a decrease of the intensity of the unshifted line is observed. The effect can be seen both in powder and in single crystals and can be ascribed to the formation of precursor clusters. From a comparison of the two cases it appears, however, that in the presence of a large number of point defects (as expected in the powder) the time scale of the correlated fluctuations is slowed down [2.104].

d) HCl-DCl – a NQR Line-Intensity Study. In the ferroelectric phase of HCl, a continuous decrease of the ^{35}Cl NQR line is observed on approaching the transition temperature from below, in a temperature range of about 0.6°C, without any detectable change of linewidth or T_1 [2.105].

The EFGs and the NQR parameters are related to the reorientational critical dynamics of the HCl dipoles. At the transition, when the dipole reorientation is no longer biased along the electric polarization, a large shift of the quadrupole frequency has to be expected. The qualitative conclusion drawn from the behavior of the line intensity and of other NQR quantities with temperature is that two phases are present: one for which order parameter is almost zero (paraelectric phase), which cannot be detected, and a second one with order parameter almost equal to one, with a characteristic correlation time for the dipole reorientation of the order of 10^{-9} s (Panel 2.46).

From the constancy of $\delta\nu$, the trivial rounding of the transition due to impurities and/or defects, which do not couple to the order parameter, is ruled out. The same behavior of the NQR quantities is observed in a mixed crystal HCl-DCl. Thus the effect cannot be simply ascribed to a local freezing of the naturally abundant deuterium atoms (coupled to the critical dynamics, with different tunneling). An intrinsic mechanism of heterophase fluctuations is the probable explanation, whereby the negligible change of shape and volume at the transition, with a small coherence strain energy, is a possible favoring effect.

e) NbSe$_2$ – a Line-Broadening Study on a CDW System. In the dichalcogenide layered compound 2H-NbSe$_2$, an incommensurate CDW phase sets in at $T_0 = 33$ K (Panel 2.38). ^{93}Nb linewidth and relaxation measurements for $T \rightarrow T_0$ have been performed in order to investigate possible precursor effects [2.106]. A marked broadening, both of the central transition line and of the satellite lines, is observed (Panel 2.47). The broadening could be due, in principle, either to an extrinsic mechanism consisting of the formation of CDW regions pinned by impurities or to an intrinsic mechanism related to dynamic CDW fluctuations causing a second-order transition-type CP. Since no critical effects can be observed either in the spin-spin relaxation rate (homogeneous broadening) or on the spin-lattice relaxation rate $T_1^{-1} \simeq \sum_{\bar{q}} S(\bar{q}, \omega_L) \simeq |\omega_Q|^2 \tau_c$, the conclusion is in favor of an inhomogeneous pretransitional broadening of extrinsic origin.

f) KDA – a Rotation Pattern Study of Quadrupole Perturbed NMR Spectra. KH$_2$AsO$_4$ (KDA) represents one of the first crystals in which pretransitional clusters have been detected by NMR–NQR. Above the paraelectric-ferroelectric transition at $T_c = 96$ K, besides the ^{75}As quadrupole-perturbed spectrum pertaining to the paraelectric phase, additional anomalous lines have been observed [2.107]. While the

---

PANEL 2.46 _____

NQR Evidence for Pretransitional Clusters

- In K$_2$OsCl$_6$, at $T \simeq 43.7$ K, one observes the appearance of a broad line shifted by about 6 kHz (Fig. 2.46a). This line corresponds to a short-range order parameter (angle of rotation of the OsCl$_6$ octahedra) of about 1.5°.
- In the HCl crystal (Fig. 2.46b), when the dipoles reorient, the V_{zz} axis of the EFG tensor also reorients.
- The reorientation of V_{zz} directly affects the NQR parameters:

$$\nu_R = (\nu_Q/2)(1 + 3\langle s \rangle)^{1/2} \quad ,$$

where $\langle s \rangle$ is the order parameter, and $T_1^{-1} = \frac{3}{2}\pi^2 \nu_0^2 J(\omega_R)$, where $J(\omega)$ is the spectral density for the motion. The intensity I traces the number of dipoles belonging to the ferroelectric clusters (Fig. 2.46c). The lifetime of the clusters is at least $(\nu_Q/2)^{-1}$ i.e. 10^{-7} s or longer.

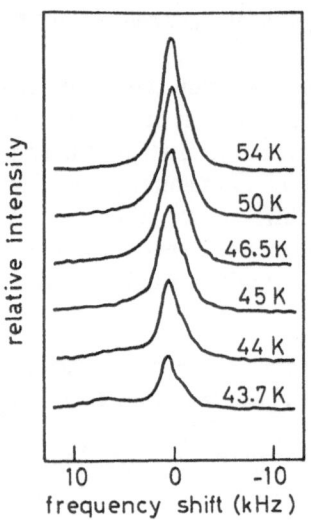

Fig. 2.46a. ^{35}Cl NQR spectra in K_2OsCl_6 showing the decrease with temperature of the main line and the appearance of a second line on approaching the transition [2.104]

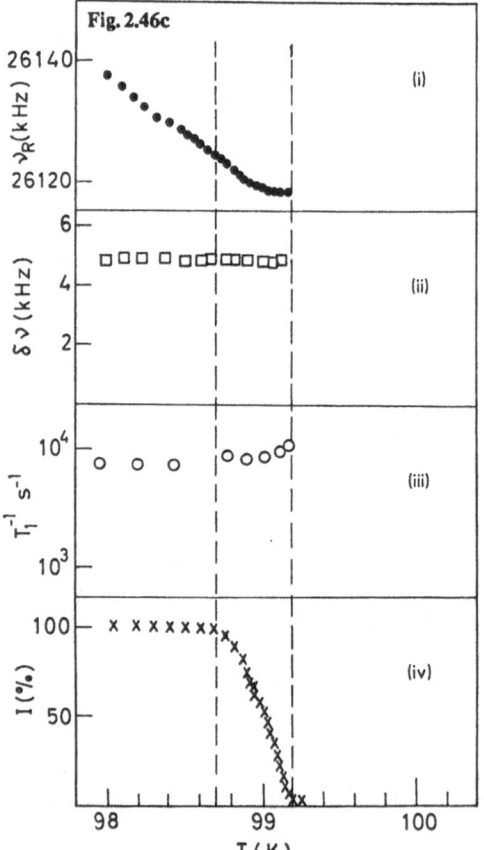

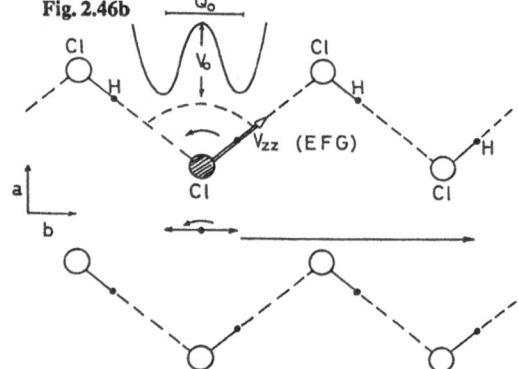

Fig. 2.46b. Schematic structure of HCl crystal evidencing the reorientation of the dipole and of V_{zz}

Fig. 2.46c. Behavior of ^{35}Cl NQR quantities in HCl on approaching the ferroelectric-paraelectric transition: (*i*) resonance frequency; (*ii*) linewidth; (*iii*) relaxation rate; (*iv*) intensity of the line

paraelectric lines do not have any angular dependence, upon rotation around the polar axis perpendicular to H_0 the additional lines are expected to display the ordinary rotation pattern. The early measurements [2.107] were performed in a low magnetic field, where the Zeeman interaction is comparable to the quadrupole interaction, and did not allow a complete rotation pattern to be carried out. The incomplete rotation patterns did not permit an unambiguous EFG assignment and, on the basis of an interpretative model, the extra lines were attributed to nuclei residing, for short times, in mobile fully polarized ferroelectric clusters. A detailed study of T_1, $T_{1\varrho}$ and T_2 would be needed to establish whether a progressive slowing-down of dynamic clusters takes place on cooling towards T_c.

More recent measurements [2.108] in higher fields, indicate the presence of four chemically equivalent EFG tensors, corresponding to the two crystallographic As sites, each in two regions of opposite polarization (Panel 2.47). The EFG assignment allowed the authors [2.108] to conclude that the extra lines arise from nuclei residing in regions of the crystal where the EFG tensor has the same symmetry as the ferroelectric phase, but with a reduced strength of V_{zz}. The simplest explanation consists in assuming the presence of quasi-static ($\tau \geq 10^{-3}$ s) partially polarized clusters of extrinsic origin; see also Chap. 1.

PANEL 2.47 ⎯⎯⎯⎯⎯⎯⎯⎯⎯⎯⎯⎯⎯⎯⎯⎯⎯⎯⎯⎯⎯⎯⎯⎯⎯⎯⎯⎯⎯⎯⎯⎯⎯

NMR Evidence for Dynamical Pretransitional Effects

- In NbSe$_2$ the broadening of the $\left(-\frac{3}{2} \leftrightarrow -\frac{1}{2}\right)$ ^{93}Nb satellite line is sensitive to the EFG distribution to first order, as generated by the onset of CDW modulations of variable amplitudes close to T_c, which are pinned at the sites of impurities and/or defects (Fig. 2.47a).

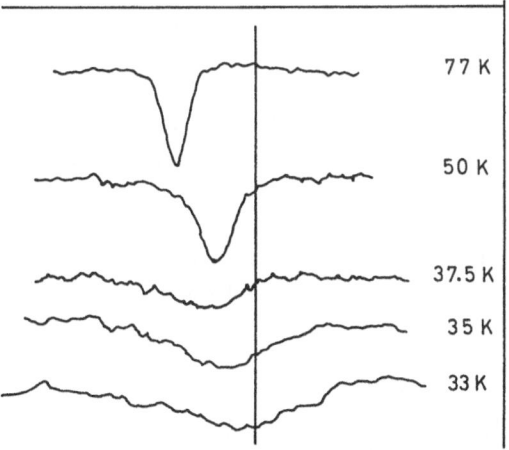
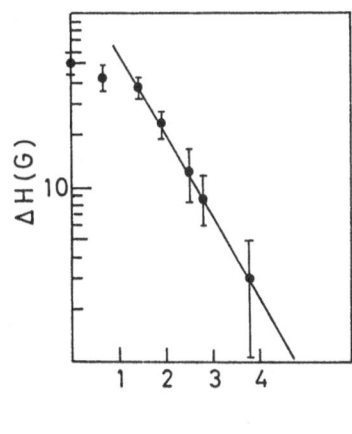

Fig. 2.47a. Broadening of a ^{93}Nb satellite line observed in NbSe$_2$ and plot of the width as a function of $\ln(T - 33)$ [2.106]

- In K_2AsPO_4, for the $I = \frac{3}{2}$ ^{75}As NMR quadrupole-perturbed spectrum the second-order shift of the central line for $\theta = \pi/2$ is:

$$\Delta\nu_{1/2}^{(2)} = -\left(\nu_Q^2/2\nu_L\right)\left[\tfrac{3}{8} + \tfrac{1}{3}\eta^2 - \frac{\eta}{4}\cos 2\phi - \tfrac{3}{8}\eta^2\cos^2 2\phi\right] \quad .$$

The paraelectric line (Fig. 2.47b) is angle independent since $\eta = 0$. The four ferroelectric-type lines have $\eta = 0.35$ and $A_{1,2}$ and $B_{1,2}$ correspond to the minor principal axes rotated by $\pm 26°$ and $\pm 64°$, respectively, with respect to the a-axis.

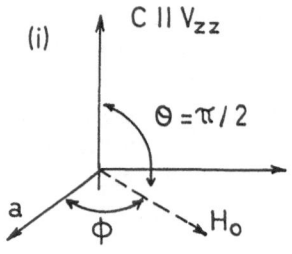

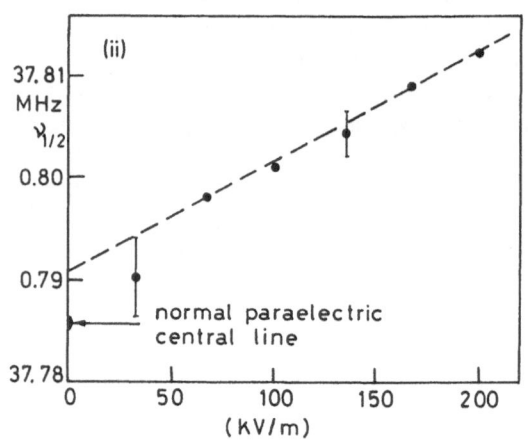

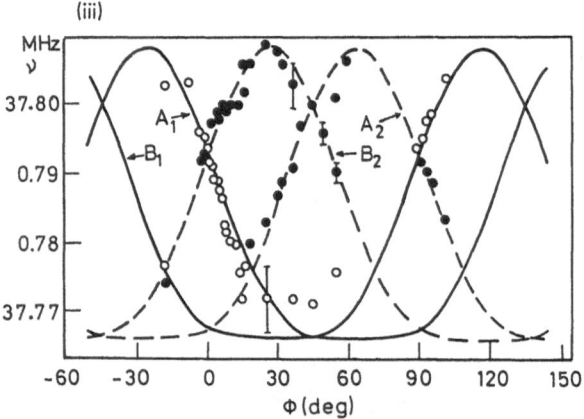

Fig. 2.47b.
Rotation pattern (*i*) of the ^{75}As NMR central line and effects of the external electric field (*ii*). The A_1, B_1 and A_2, B_2 groups of lines (*iii*) are enhanced (and shifted) in opposite directions by an external electric field [2.108]

Acknowledgments. Special thanks go to Piera Priori and Viviana Minassi for their contribution to the preparation of the manuscript. P. Orlandi is gratefully thanked for his technical help and for the graphics in the panels.

References

2.1   L.D. Landau, E.M. Lifshitz: *Statistical Physics* (Addison-Wesley, Reading, Mass. 1969)

2.2   H.E. Stanley: *Introduction to Phase Transitions and Critical Phenomena* (Clarendon Press, Oxford 1971)

2.3   H. Thomas: In *Structural and Soft Modes*, ed. by E.J. Samuelsen, E. Andersen, J. Feder (Universitetsforlaget, Oslo 1971)

2.4   M.E. Lines, A.M. Glass: *Principles and Applications of Ferroelectrics and Related Materials* (Clarendon Press, Oxford 1977)

2.5   B.I. Halperin, P.C. Hohenberg: Rev. Mod. Phys. **49**, 435 (1977)

2.6   K.A. Müller, A. Rigamonti (Eds.): *Local Properties at Phase Transitions* (North-Holland, Amsterdam 1976)

2.7   R. Balian, R. Maynard, G. Toulouse (Eds.): *Ill-condensed Matter* (North-Holland, Amsterdam 1983)

2.8   S.F. Edwards, P.W. Anderson: J. Physics F **5**, 965 (1975)

2.9   G. Toulouse: Helv. Physica Acta **57**, 459 (1984)

2.10 H. Sompolinski: Phys. Rev. Lett. **47**, 935 (1981)

2.11 See P.W. Anderson in [2.7]

2.12 J.A. Krumhansl, J.R. Schrieffer: Phys. Rev. B **11**, 3535 (1975)

2.13 S. Aubry: J. Chem. Phys. **62**, 3217 (1975)

2.14 J. Villain: In *Ordering in Strongly Fluctuating Condensed Matter Systems*, ed. by T. Riste (Plenum Press, New York 1980)

2.15 B.I. Halperin, C.M. Varma: Phys. Rev. B **14**, 4030 (1976)

2.16 K.H. Hoch, H. Thomas: Z. Phys. B **27**, 267 (1977);
see also K.H. Hock, R. Schafer, H. Thomas: Z. Phys. B **36**, 151 (1973)

2.17 See, for example: T. Schneider, A. Stoll: Phys. Rev. Lett. **31**, 1254 (1973);
S. Aubry, R. Pick: Ferroelectrics **8**, 471 (1974)

2.18 C.M. Varma: Phys. Rev. B **14**, 244 (1976)

2.19 H. Beck: J. Phys. C **9**, 33 (1976)

2.20 H. Tani, M. Takemura: J. Phys. Soc. Jpn. **30**, 328 (1971);
R.A. Cowley, G.J. Coombs, R.S. Katiyar, J.F. Rayan, J.F. Scott: J. Phys. C **4**, L 203 (1971)

2.21 G. Shirane, J.D. Axe: Phys. Rev. Lett. **27**, 1803 (1971);
S.M. Shapiro, J.D. Axe, G. Shirane, T. Riste: Phys. Rev. B **6**, 4332 (1972);
F. Schwabl: Z. Phys. **254**, 7 (1972)

2.22 K.A. Müller: In *Dynamical Critical Phenomena and Related Topics*, Lecture Notes in Physics, Vol. 104, ed. by C.P. Enz (Springer, Berlin, Heidelberg 1979) p. 210;
R. Blinc: Ferroelectrics **20**, 121 (1978);
A.D. Bruce: In *Solitons and Condensed Matter Physics*, Springer Ser. Solid-State Sci., Vol. 8, ed. by A.R. Bishop, T. Schneider (Springer, Berlin, Heidelberg 1981) p. 116

2.23 H.E. Cook: Phys. Rev. B **15**, 1477 (1977)

2.24 For a comprehensive review of optical studies, see P.A. Fleury, K.B. Lyons: "Optical Studies of Structural Phase Transitions" in *Structural Phase Transitions*, ed. by K.A. Müller, H. Thomas (Springer, Berlin, Heidelberg 1981)

2.25 F. Schwabl: Z. Phys. **254**, 7 (1972)

2.26 P.F. Meier: Solid State Commun. **13**, 967 (1973)

2.27 S.M. Shapiro, J.D. Axe, G. Shirane, T. Riste: Phys. Rev. B **6**, 4332 (1972)

2.28 J. Topler, B. Alefeld, A. Kollmar: Phys. Lett. **51A**, 297 (1975) and references therein;
C.N.W. Darlington, W.J. Fitzgerald, D.A. Connor: Phys. Lett. **54A**, 35 (1974)

2.29 For a comprehensive analysis of neutron scattering results in the light of the CP and precursor clusters effects, see: A.D. Bruce, R.A. Cowley: Adv. Phys. **29** (1980)

2.30 I. Hatta, M. Matsuda, S. Savada: J. Phys. C **7**, L 299 (1974)

2.31 R. Blinc, B. Zeks: *Soft Modes in Ferroelectrics and Antiferroelectrics* (North-Holland, Amsterdam 1974)

2.32 R.A. Cowley, A.D. Bruce: J. Phys. C11, 3577 (1978);
A.D. Bruce, R.A. Cowley, A.F. Murray: J. Phys. C11, 3591 (1978)

2.33 R. Blinc: Physics Reports 79, 331 (1982); a more recent review on fundamentals and materials is R. Blinc, A.P. Levaniuk (Eds.) *Incommensurate Phases in Dielectrics* (North-Holland, Amsterdam 1986)

2.34 A. Abragam: *Principles of Nuclear Magnetism* (Clarendon Press, Oxford 1961)

2.35 E. Fukushima, G. Roeder: *Experimental Pulse NMR: A Nuts and Bolts Approach* (Addison-Wesley, New York 1981)

2.36 M. Goldman: *Spin Temperature and Nuclear Magnetic Resonance in Solids* (Clarendon Press, Oxford 1970)

2.37 For a reformulation of the relaxation processes applied to phase transitions and critical phenomena, see F. Borsa, A. Rigamonti: In *Magnetic Resonance of Phase Transitions*, ed. by F.J. Owens, C.P. Poole, H.A. Farach (Academic, New York 1979)

2.38 A. Rigamonti: Adv. in Phys. 33, 115 (1984)

2.39 K.A. Müller: In *Local Properties at Phase Transitions*, ed. by K.A. Müller, A. Rigamonti (North-Holland, Amsterdam 1976)

2.40 F. Borsa: Phys. Rev. 7, 915 (1973)

2.41 S.V. Bhat, P.P. Mahendroo, A. Rigamonti: Phys. Rev. B20, 1812 (1979)

2.42 G.D'Ariano, S. Aldrovandi, A. Rigamonti: Phys. Rev. B25, 7044 (1983)

2.43 A.G. Brown, R.L. Armstrong, K.R. Jeffrey: Phys. Rev. B8, 121 (1973)

2.44 R. Kind, J. Roos: Phys. Rev. 13, 45 (1976)

2.45 M. Mehring, J.D. Becker: Phys. Rev. Lett. 47, 366 (1981)

2.46 R. Blinc, B. Lozar, B. Topic, S. Zumer: J. Phys. C16, 5053 (1983)

2.47 T. Guillon, M.S. Conradi, A. Rigamonti: Phys. Rev. B31, 4388 (1985)

2.48 R. Blinc: In *Magnetic Resonance of Phase Transitions*, ed. by F.J. Owens, C.P. Poole, H.A. Farach (Academic, New York 1979)

2.49 Y.H. Seo, J. Pelzl, C. Dimitropouls: Z. Naturforschung 41a, 311 (1986);
J. Pelzl, V. Waschz, Y.M. Seo, C. Dimitropouls: J. Molec. Structure 11, 363 (1983) and private communication by J. Pelzl

2.50 F. Borsa, D.J. Benard, W.C. Walker, A. Baviera: Phys. Rev. B15, 84 (1977)

2.51 E. Courtens: Helv. Phys. Acta 56, 705 (1983);
see also E. Courtens: Jpn. J. Appl. Phys. Suppl. 24, 70 (1985)

2.52 R. Blinc, D.C. Ailon, B. Gunther, S. Zumer: Phys. Rev. Lett. 57. 2826 (1986)

2.53 D. Ritz: Thesis, Lausanne (1983)

2.54 J.J. Van der Klink, S. Rod, A. Chatelain: Phys. Rev. B33, 2084 (1986);
F. Borsa, R. Lecander, A. Rigamonti: unpublished; see also [2.38]

2.55 S. Rod, F. Borsa, J.J. Van der Klink: Phys. Rev. B38, 2267 (1988)

2.56 S. Elschener, J. Petersoni: Z. Phys. B52, 37 (1983)

2.57 S. Elschner, K. Knorr, A. Loidl: Z. Phys. B61, 209 (1985)

2.58 S. Elschner, J. Petersson: J. Phys. C19, 3373 (1986)

2.59 N.S. Sullivan, J.M. Vaissiere: Phys. Rev. Lett. 51, 658 (1983)
See also: N.S. Sullivan, D. Esteve, M. Devoret: J. Phys. C15, 4895 (1982)

2.60 N.S. Sullivan, C.M. Edwards, Y. Lin, D. Zhou: International Conference on Quantum Fluids, Banff, Canada (1986)

2.61 J.J. Van der Klink, D. Rytz, F. Borsa, U.T. Hochli: Phys. Rev. B27, 89 (1983)

2.62 J.J. Van der Klink, F. Borsa: Phys. Rev. B30, 52 (1984)

2.63 L.L. Chase, E. Lee, R.L. Prater, L.A. Boatner: Phys. Rev. B26, 2759 (1982)

2.64 M. Maglione, S. Rod, U. Hochli: Europhys. Lett. 4, 631 (1987)

2.65 U.T. Hochli, A. Rigamonti: J. Phys. C16, 6321 (1983)

2.66 J.J. Van der Klink, D. Rytz: Phys. Rev. B27, 4471 (1983)

2.67 A. Rigamonti, S. Torre: Phys. Rev. B33, 2024 (1986);
see also S. Torre, A. Rigamonti: Phys. Rev. B36, 8274 (1987)

2.68 A. Rigamonti, S. Torre: Solid State Commun. 56, 619 (1985)

2.69 L. Lundgren, P. Svdeinder, O. Beckman: J. Magn. Magn. Mater. 31–34, 1349 (1983)

2.70   O. Kanert, R. Kuchler, M. Mali: J. de Phys. Colloq. **41**, C6–404 (1980)

2.71   J. Slak, R. Kind, R. Blinc, E. Courtens, S. Zumer: Phys. Rev. B **30**, 85 (1984)

2.72   R. Kind, O. Liechti, R. Bruschweiler, J. Dolinsek, R. Blinc: Phys. Rev. B **36**, 13 (1987); see also
       W.T. Sobol, J.G. Cameron, M.M. Pintar, R. Blinc: Phys. Rev. B **35**, 7299 (1987)

2.73   P. Calvani, H. Glattli: J. Chem. Phys. **83**, 1822 (1986)

2.74   M.A. Doverspike, Meng-Chou Wu, M.S. Conradi: Phys. Rev. Lett. **56**, 2284 (1986)

2.75   M. Crowley, J. Brookeman, A. Rigamonti: Phys. Rev. B **28**, 5184 (1983)

2.76   J. Szeftel, H. Allou: J. non-Crystalline Solids **29**, 253 (1978)

2.77   R. Blinc, A.P. Levanyuk (eds.): *Incommensurate Phases in Dielectrics* (North-Holland, Amsterdam
       1986)

2.78   J. Dolinsek, S. Zumer, R. Blinc: *Proc. 22nd Colloque* (Ampere, Zurich 1984)

2.79   R. Ambrosetti, N. Angelone, A. Colligiani, A. Rigamonti: Phys. Rev. B **15**, 4318 (1977)

2.80   I.P. Alexandrova, R. Blinc, B. Topic, S. Zumer, A. Rigamonti: Phys. Status Solidi A **61**, 95 (1980)

2.81   W. Buchheit, G. Herth, J. Petersson: Solid State Commun. **40**, 411 (1981)

2.82   R. Blinc, B. Lozar, F. Milia, R. Kind: J. Phys. C **17**, 241 (1984) and references therein

2.83   J. Petersson, E. Schneider: Z. Phys. B **61**, 33 (1985)

2.84   A.H. Kabiza, M. Pezeril, J. Emery, J.C. Fayet: Ferroelectrics **53**, 261 (1984)

2.85   C. Berthier, D. Jerome, P. Molinie: J. Phys. C **11**, 798 (1978) and references therein

2.86   F. Borsa, D.R. Torgeson, H.R. Shanks: Phys. Rev. B **15**, 4576 (1977);
       B.H. Suits, C.P. Slichter: Phys. Rev. B **29**, 41 (1984)

2.87   S.W. Vada, R. Aoki, O. Fujita: J. Phys. F **14**, 1515 (1984)

2.88   J.H. Ross, Jr., Zhiyne Wang, C.P. Slichter: Phys. Rev. Lett. **56**, 663 (1986)

2.89   P. Segransand, A. Janossi, C. Berthier, J. Marcus, P. Buton: Phys. Rev. Lett. **56**, 1854 (1986)

2.90   S. Zumer, R. Blinc: J. Phys. C **14**, 465 (1981)

2.91   A.D. Bruce, R.A. Cowley: J. Phys. C **11**, 3609 (1978)

2.92   R. Blinc, D.C. Ailion, J. Dolinsek, S. Zumer: Phys. Rev. Lett. **54**, 79 (1985)

2.93   R. Blinc, F. Milia, V. Rutar, S. Zumer: Phys. Rev. Lett. **48**, 47 (1982)

2.94   S.B. Liu, M.S. Conradi: Phys. Rev. Lett. **54**, 1287 (1985)

2.95   M.S. Dresselhaus, G. Dresselhaus: Adv. Phys. **30**, 139 (1981)

2.96   F. Borsa, M. Corti, A. Rigamonti, S. Torre: Phys. Rev. Lett. **53**, 2102 (1984)

2.97   M. Suzukiand, H. Suematzu: J. Phys. Soc. Jpn. **52**, 2761 (1983)

2.98   R. Blinc, J. Slak, F.C. Barreto, A.S.T. Pires: Phys. Rev. Lett. **42**, 1000 (1979)

2.99   A. Avogadro, G. Bonera, A. Rigamonti: J. Magn. Resonance **20**, 399 (1975)

2.100 A. Trokiner, P.P. Man, H. Theveneau, P. Papon: Solid State Commun. **55**, 929 (1985)

2.101 A. Bulou, H. Theveneau, A. Trokiner, P. Papon: J. Phys. Lett. **40**, L 277 (1979)

2.102 A. Trokiner, H. Theveneau, P. Papon: *Proc. of the XXIII Congress Ampere (Rome 1986)*, ed. by B.
       Maraviglia, F. De Luca, R. Campanella, p. 222; see also
       A. Trokiner, N. Dahan, J.L. Miquel, H. Theveneau, P. Papon: Physica **139**, 319 (1986)

2.103 M. Mehring, D. Suwelack: Phys. Rev. Lett. **42** (1979)

2.104 R.L. Armstrong, M. Ramia: J. Phys. C **18**, 2977 (1975)

2.105 J. Brookeman, A. Rigamonti: Phys. Rev. B **24**, 4925 (1981)

2.106 C. Berthier, D. Jerome, P. Molinie: J. Phys. **11**, 797 (1975)

2.107 G.J. Adriaenssens: Phys. Rev. B **12**, 5116 (1975); see also
       R. Blinc, J.L. Bjorkstam: Phys. Rev. Lett. **23**, 788 (1969)

2.108 M. Mali, J. Roos, E. Courtens, K.A. Muller: Ferroelectrics **53**, 215 (1984)

Subject Index

177